MAKING ONLINE NEWS
VOLUME 2

Steve Jones
General Editor

Vol. 67

PETER LANG
New York • Washington, D.C./Baltimore • Bern
Frankfurt • Berlin • Brussels • Vienna • Oxford

MAKING ONLINE NEWS
VOLUME 2

Newsroom Ethnographies in the Second Decade of Internet Journalism

Edited by David Domingo & Chris Paterson

PETER LANG
New York • Washington, D.C./Baltimore • Bern
Frankfurt • Berlin • Brussels • Vienna • Oxford

The Library of Congress has catalogued the first volume as follows:

Making online news: the ethnography of new media production /
edited by Chris Paterson, David Domingo.
p. cm.—(Digital formations; vol. 49)
Includes bibliographical references and index.
1. Online journalism. I. Paterson, Chris. II. Domingo, David.
PN4784.O62M35   302.231—dc22   2008010857
ISBN 978-1-4331-0214-1 (volume 1, hardcover)
ISBN 978-1-4331-0213-4 (volume 1, paperback)
ISBN 978-1-4331-1065-8 (volume 2, hardcover)
ISBN 978-1-4331-1064-1 (volume 2, paperback)
ISSN 1526-3169

Bibliographic information published by **Die Deutsche Nationalbibliothek.**
Die Deutsche Nationalbibliothek lists this publication in the "Deutsche
Nationalbibliografie"; detailed bibliographic data is available
on the Internet at http://dnb.d-nb.de/.

The cover photograph of the Associated Press New York newsroom
is by Richard Drew, and was provided courtesy of the Associated Press.

The paper in this book meets the guidelines for permanence and durability
of the Committee on Production Guidelines for Book Longevity
of the Council of Library Resources.

© 2011 Peter Lang Publishing, Inc., New York
29 Broadway, 18th floor, New York, NY 10006
www.peterlang.com

Printed in the United States of America

Contents

PART TWO: NEWS REDEFINED

PART THREE: BEYOND THE NEWSROOM

A Pedagogy of Online News Sociology
Teaching with *Making Online News*

Chris Paterson and David Domingo

This second volume of *Making Online News* updates and extends the unique international collection of online newsroom research provided by the first volume in 2008. Together, we hope for the two books to provide a versatile teaching tool for the ever-increasing number of educators seeking to explain change in journalism and prepare students to be journalists in an environment vastly different from that described by the standard canon of journalism theory and how-to guides. Allowing contemporary research to shape pedagogy is often a challenge, especially in a fast-changing field where comprehensive and authoritative research remains sparse. The challenge is compounded by an ever more pervasive assembly-line approach to journalism education which marginalizes critical reflection and engagement with research. The two volumes of *Making Online News* take educators and students into over thirty newsrooms around the world in the context of cutting-edge analysis of the changes underway there.

CHRIS PATERSON AND DAVID DOMINGO

In a review of textbooks on convergent journalism, Gilmour & Quanbeck (2010) shared their frustration that the texts they came across were narrowly focused on technical skills and acritically followed job-market imperatives—typically preaching the mantra that "multiskilled journalists" are what companies need. Over all, what most worried these news production instructors was that many textbooks "ignore the societal obligations of journalism":

> The absence of focus on the democratic principles of journalism in some of the textbooks debases the books. Authors who do not mention or explain to students the democratic principles of journalism or urge students to take up the long-standing democratic principles of journalism merely write a how-to manual. The practice of journalism is more than a formula to be followed. (Gilmour & Quanbeck, 2010, p. 338)

In other words, although industry managers occasionally suggest otherwise, the first obligation of journalism educators is to foster intellectual engagement and a deep sense of social responsibility in students of journalism. Both theory and skills classes have to convey these principles. Journalism does not exist in a vacuum, and a young journalist with little concept of journalistic obligation, of the social context of the field, and of the possibilities, limitations, and evolution of the latest forms of journalistic practice, may well do society, and a news organization, more harm than good. Online journalism textbooks tend to focus on the *potential* storytelling opportunities of the Internet and the practical skills to produce a new kind of news for the Web. Learning about *actual* developments in actual newsrooms from a critical perspective is essential, and in this *Making Online News* can help.

In order to empower future journalists to work in the evolving digital news environment and to make informed decisions to shape news production towards a more socially responsible, relevant, journalism, they need to be aware of the organizational constraints, the dynamics of change, and the redefinition of roles and values happening in newsrooms. The ethnographic research data in the two volumes of *Making Online News* provide the opportunity to turn hands-on courses into fruitful debates contrasting what students are experimenting with in their online news production labs and the challenges faced by professional journalists in online newsrooms around the world.

In theoretical courses on current trends in (online) journalism, the cases analyzed in the two volumes of *Making Online News* offer unique starting points to reflect on the convulsive times media face. We believe these books are a good tool to foster the "critical-thinking skills" that Gilmour and Quanbeck (2010, p. 339) deemed essential in journalism education. Below educators will find some recommendations on how to use the chapters in the first and second volumes of the book to teach different themes. You may propose that students read all the chapters on a theme and, through class discussion or written work, contrast the perspectives of the different cases; or that students match selected readings for the themes below to supplement your syllabus. We would like to hear about how you use Making Online News in your teaching. Please join the conversation with us at www.makingonlinenews.net.

A Suggestion of Clustered Readings from *Making Online News*

Convergence

A good starting point is Paulussen et al. (vol. 2); this is a chapter which offers a thorough literature review of research on the factors shaping organizational change in newsrooms and discusses the theme of collaboration with two cases in Belgium. Colson and Heinderyckx (vol. 1) highlight the tensions that usually arise in the process of convergence between the vision of the management and the perspective of the reporters. Bechmann (vol. 2) provides a critical analysis of the implications of one of the most celebrated elements of newsroom integration, the "superdesk" or multimedia hub, showing that there are other factors defining convergent news production dynamics besides the mere physical proximity of the different teams. For PhD seminars, Singer (vol. 1) offers a thoughtful reflection on the benefit of an ethnographic methodology in convergence research.

Professional Identity

The culture clashes between the print and online journalistic cultures reveal the process of (re)definition of the professional identity of (online) journalists. This can be illustrated by three cases with an international perspective. Cawley (vol. 1) depicts the dynamics of news production in the online newsroom of *The Irish Times* and how it fits in the established newspaper culture. García (vol. 1) explicitly addresses the frustration of the online staff at the Argentinean *Clarín* as they are treated like "second class" citizens by their print counterparts. Robinson's (vol. 2) contribution is a fascinating tale of what happens when radical decisions such as stopping the presses and making just online news are taken at a US regional newspaper, *The Capital Times*.

Online News Values and Formats

The shape of online news is embedded in the practices of newsrooms. Four chapters, in particular, provide a comprehensive picture of how news for the Web is defined and produced in actual newsrooms. Both Domingo (chapter 7 in vol. 1) and Møller Hartley (vol. 2) discuss the most central product of online newsrooms: breaking news. Domingo focuses on the organizational decisions leading to privileging immediacy as the main news value, while Møller Hartley revisits Tuchman's

(1978) typology of news to understand how online newsrooms deal with different
kinds of stories. Van Dam (vol. 2) describes how two different models of newsroom
deal differently with one of the biggest news events: a US presidential campaign.
Another perspective which might be included in this thread is Steensen's (vol. 2)
discussion of feature reporting and how genres are shaped in online newsrooms.
The decisions regarding what is *news* and how it is told, as described in these chapters, might usefully shape classroom discussion of the potential features of online
storytelling and the factors constraining those features.

News Agency Wires and Online Journalism

Research shows the prominent role of news agency material in online news
production. To illustrate this point and the production routines surrounding it,
the studies of Domingo (chapter 7 in vol. 1) and Quandt (vol. 1) offer insights on
the tasks of journalists and their handling of wire copy. As a complementary perspective, the chapter by Paterson (vol. 2) on the modern evolution of the international
news agencies highlights how the material reaching the newsrooms is produced.
Firdaus (vol. 2) further shows how marketing logics are at the core of the online
activities of national news agencies. These readings offer a starting point to reflect
on the homogenization of news in online journalism and perhaps celebrate all the
more those online news organizations managing genuinely original reporting and
investigation.

Multimedia Production

The chapters dealing with this aspect most explicitly are those of Brannon (vol.
1) and Usher (vol. 2). Both researched online newsrooms among US broadcasters
with almost a decade between them, which provides a useful element of historical
contrast in reading the chapters. Brannon focuses on the technical and organizational
constraints to the development of multimedia storytelling in the early days of
Internet news. Usher describes how a very contemporary national radio program
is adopting online tools such as blogs and podcasts to support their main activity,
while exploring new ways to present their journalism.

Blogs and Journalism

The blurring boundaries between professional journalism and the discussions
led by bloggers outside mainstream media have many implications for journalistic
identity and the dynamics of news production and diffusion. Bruns (vol. 1) offered
a thorough theoretical framework for the emerging rules of a more plural online
news arena. Lowrey and Latta (vol. 1) demonstrated how bloggers tend to replicate
routines of professional journalism when they get serious about their postings. And

Anderson (vol. 2) bridges the two worlds, the newsroom and the blogosphere, analyzing a local news ecosystem and the attitudes of both journalists and bloggers in this open context, concluding that the newsroom still holds a central position in defining news.

Participatory Journalism

Building a relationship with the active audience—those users that are willing to produce content for professional news websites—has become one of the central strategies of online journalism. Williams et al. (vol. 2) discuss the diverse attitudes and management strategies at the BBC, showing how there is not a single solution even within the same institution. Paulussen, Bechmann and Robinson (all in vol. 2) also offer evidence of newsroom strategies for audience participation in portions of their chapters. All of them highlight the resistance of journalists to redefine their identity in the new context and explain how user participation ends up being accommodated into traditional news production practices.

Journalism under Pressure

Several chapters discuss situations where journalism is under more pressure than usual, facing a context of state control of the media. Lagerkvist (vol. 1) presents the case of China, Mabweazara (vol. 2) introduces Zimbabwe, and Firdaus (vol. 2) discusses Malaysia. While each focuses on different aspects of online journalism, they can be discussed together to illustrate how the Internet develops as a space with the potential to challenge authoritarian control as well as the potential to *exacerbate* the limitations on journalism in such circumstances.

Methodology and Epistemology of Online Journalism Research

All the chapters deal with aspects of ethnographic research design and theoretical frameworks for the analysis of online journalism, but some are very explicit about this and so might form a useful group of readings for graduate research seminars. The introduction by Paterson and the chapter by Singer in the first edition are overviews of the benefits of ethnographic methods in newsroom studies. Puijk (vol. 1) discusses more practical aspects of conducting observation in digital newsrooms. Domingo (chapter 1 in vol. 1) provides an overview of the epistemological principles of online newsroom ethnographies and the traditions that have inspired their theoretical frameworks. Anderson (vol. 2) challenges the definition of the object of study and proposes problematizing the newsroom itself as the scenario of observation, expanding it to the whole news ecosystem to include bloggers and other institutions interacting with the journalists in creating the news.

The Centrality of Online
Journalism Today (and Tomorrow)

David Domingo

"We will stop printing the *New York Times* sometime in the future, date TBD."
Arthur Sulzberger Jr, the publisher of one of the most influential newspapers in
the world, joked about the end of an era in a conference in London in September
2010. He was presenting the new paywall strategy of their website operation, but
the audience was more eager to know if the print edition would become history
any time soon. Why does this matter when we discuss the present and future of
online journalism? The development of online journalism does not happen in isolation. If the first edition of *Making Online News* showed that the first decade of
online news history was determined by the tensions between the possibilities of the
Internet and the conservative attitudes of the newsrooms, which naturally tamed
the *radical potential* of the new technology (Winston, 1998), the second decade is
reassuring in that online journalism has become a central element in the struggle
of the media industry to reinvent itself in order to adapt to structural social changes
and overcome the slow demise of its traditional business model.

The Internet has taken a central position in current debates about the future
of journalism.[1] The hesitant exploration of newsroom convergence—disparate
approaches to and strategies regarding the production of print, broadcast and online
news are multiplying (García Avilés et al., 2009)—is the most tangible evidence
that the evolution of online journalism cannot be untangled from the development
of the rest of the profession. And the enthusiastic but careful embracing of audience

participation in news websites, with newsrooms making sure that their gatekeeping role is not challenged (Hermida and Thurman, 2008; Singer et al., 2011), shows how the professional culture of journalism actively negotiates its role and legitimacy in an evolving communicative space. Online journalism is here to stay: that is already taken for granted. Its shape, though, is still very much under construction, a long, open-ended process that is not going to be placid. Many tensions still pull the future of online journalism in different directions, and the quality of news and its role in democracy may be at stake.

Net-native journalistic projects[2] are a good observatory of the intrinsic tensions in the evolution of online media. In many of them there is a vindication of old journalistic values such as beat and investigative reporting and a stronger connection to the community they serve. At the same time, there is a celebration of opinionated writing and user-generated content that stretches the orthodox definitions of journalism. While online journalism has been recognized already with Pulitzer Prizes in the USA, most of the news stories produced online are mere clones of agency wire copy (Paterson, 2007). And there is an increasing pressure of the market logics as online media revenues are not compensating for the losses of the rest of the industry. This pressure is being addressed with divergent strategies: nonprofit initiatives based on donations explore organizational and legal solutions to find a sustainable alternative model (Downie & Schudson, 2009) side by side with the downsizing of for-profit newsrooms to save structural costs (Deuze, 2007). Also, some voices in the USA suggest that public subsidies could be one of the few ways to guarantee the survival of socially committed journalism (McChesney & Nichols, 2010), especially at a local level (Downie & Schudson, 2009), where the disappearance of professional news organizations is not compensated by citizen journalism initiatives that are too fragile and lack resources for comprehensive news production (Duffy et al., 2010). In Europe, with a long tradition of state-funded media, the validity of this approach is also vindicated, but its proponents acknowledge the need to reconsider traditional public broadcasting models in the online context (Lowe, 2010). In the meantime, traditional media companies are tempted to explore (again) ways to charge for content on the Web—fighting against the ghost of the "original sin" of online journalism—and on the pay-friendly tablets and portable devices; but few are willing to develop products that stand out from mediocre and widely available free news. . Only the leaders in each market seem to capitalize with advertising revenue their investments in excellent online journalism.

The second decade of online journalism is a decade of achievements and uncertainties.[3] New forms of news production have consolidated, though not necessarily as expected. Video has become a common offering of news websites, very much following the narrative traditions of television journalism. Multimedia storytelling is mainly confined to feature reporting because it requires complex production practices, but at the same time there is a tendency to standardize the narrative structures of infographics, audio slideshows and other multimedia pieces. Data-driven journalism has reopened the debate in newsrooms regarding the need for news producers to have advanced computer skills (database programming, graphic design). Ubiquitous access to the Internet through portable devices has opened new revenue hopes, but newsrooms have a difficult time finding ways to present their journalism in this tiny new medium. Social networks have become a source, a promotional space and an interaction opportunity between journalists and their publics, challenging the boundaries between the professional and the individual persona of reporters. And Internet users have become used to a wider range of information sources about current events—from specialized blogs to community websites—while traditional media brands maintain their centrality as first-stop destinations of the vast majority of news consumers online. There is more journalism than ever being produced inside and outside professional newsrooms, but that does not necessarily mean that citizens are getting more or better news.

The Internet has consolidated as a primary source for news in most Western countries, usually just behind television.[4] Newspaper circulation has been declining since the 1990s and is today 31.5% less than 25 years ago in the US (PEJ, 2010a) and is also in a continuous descent in Europe and Australia (WAN-IFRA, 2010). The cruel irony is that while websites now represent a big percentage of newspapers' public, the print product is still their main source of revenue, as the websites contribute less than 10% of the advertising income (WAN-IFRA, 2010). Therefore, analysts expect a long transition to online-only news products. In the meantime, the last decade has seen newsrooms shrink their editorial staff all over the world, a phenomenon accelerated by the economic crisis of 2008; this has also shuttered more than 100 newspapers in the US, some of them surviving online with a fraction of their former staff (one such move online is detailed in Chapter Three of this volume). Downie and Schudson challenge the assumption that online news can be produced with fewer journalists than newspapers: "As newspapers sharply reduce their staffs and news reporting to cut costs and survive, they also reduce their value to their readers and communities" (2009: 24). The threats to journalistic quality multiply in a scenario of converged newsrooms, where reporters tend to assume more technical tasks, have "lower wages, less job security, and more contingent labor relationships" (Deuze, 2007: 147) and work under the pressure of an agenda that prioritizes covering breaking news. If research in the first volume of *Making Online News* already highlighted this trend, time has just confirmed that immediacy is the dominant paradigm of online journalism. Boczkowski (2010) warns that mimicry is a prevalent practice in online newsrooms today, in which reporting the same news as the competitors—faster if possible—is the quality measure.

Behind the numbers there is an industry struggling to define its future while it is being carried away by the everyday rhythm of current events and the cult for immediacy (Domingo, 2008a). The centrality that convergence has conveyed to online journalism is a double-edged sword: everyone expects it to be the heir of the print press, but at the same time the debate about the shape of news on the Net is blurred into the wider debates about how to save journalism. This second volume of *Making Online News* is at the same time a vindication of the growing relevance of Internet journalism in today's mediascape and a call to the necessity of more specific research into the peculiarities of online news production—only close attention will allow us to depict the particular evolution of journalism in the new medium. The first edition of the book suggested that newsroom ethnography was a fruitful methodology to unravel the social processes of innovation that were shaping the development of online journalism. This second edition presents entirely new original ethnographies conducted mostly between 2007 and 2010 to continue to critically assess the challenges, decisions and practices in online newsrooms in different corners of the world.

New Volume, New Ethnographies

The first volume has been widely read and quoted by online journalism scholars. We are very happy that ethnography has been celebrated as contributing a "nuanced analysis" to the object of study and fostered new approaches that are based on the "rejection of any deterministic perspective," as Sue Robinson put it in the *New Media and Society* review of the book. The warm welcome by the journalism studies community of the first volume and the growing number of scholars applying ethnographic approaches in their fieldwork at online newsrooms convinced us that a second edition of the book was needed.

We soon decided that it would actually be a wholly new book. The nature of ethnographic research makes it hard to revisit previous studies. To update the chapters in volume one, authors would have needed to go back to the newsrooms and would have ended up producing new ethnographies. The existing research has a lot of value in itself as an insight at a moment in time, and our pedagogical guide to the book (see the Preface to this volume) highlights the lessons we might learn. Therefore, we decided to do an open call for chapters through the email lists of communication research associations. The response was impressive: 41 studies, completed or under development, reached our mailbox and we had a hard time selecting the twelve we are presenting in this volume. This is very good news for online journalism research: there is a boom of newsroom ethnography occurring all around the world and with very diverse theoretical approaches and specific objects of study.

What ethnographic research on online news has in common is that it investigates the tensions between technological innovations and the social context where they are adopted, while always aware of the general cultural framework (journalism at large) and the particularities of the specific settings (media organizations) where decisions are being made. In a thorough review of contemporary ethnographic research on digital media, Coleman described the virtues of the method: "To grasp more fully the broader significance of digital media, its study must involve various frames of analysis, attention to history, and the local contexts and lived experiences of digital media—a task well suited to the ethnographic enterprise" (2010: 2–3). Coleman distinguished three vast areas of inquiry: the cultural politics of digital media (the communication and reinvention of identities online); the new social groups and activities that have formed within or around digital media; and the use by society of digital media in everyday life. Online newsroom research is wisely located by Coleman in the latest area, as the internet has not created an opportunity for a new social group but rather has been adopted by the existing institution of journalism and translated into work practices that at the same time accommodate to and challenge the existing professional culture.

This second edition, then, is a new collection of online newsroom ethnographies covering current issues in the evolution of news production for the Internet: we start by discussing organizational changes, new work practices and redefined professional identities, and the constraints, tensions and resistances that digitization and convergence generate. With these evolving newsrooms as the background, attention is then devoted to online news genres and the coverage of specific events. The book ends up taking a look at the context beyond the newsroom, acknowledging the increasingly crucial role of news agencies, user participation and bloggers in the configuration of online journalism.

The first section addresses the structural changes in the organization of news work. The chapter by Steve Paulussen and colleagues summarizes the factors that shape the development of collaborative initiatives across media and with citizens, two of the key trends in recent years. Anja Bechmann challenges through network analyses the effectiveness of the *superdesk* strategy that many integrated newsrooms have set up to coordinate their cross-media efforts. The work of Sue Robinson transports us to a newspaper newsroom that made the radical move to online-only, and discusses how journalists negotiated their new identity while trying to retain their core values. Nikki Usher provides the usually overlooked perspective of a radio broadcaster and shows how online innovations such as podcasts and weblogs were naturalized to fit the logics of the existing production culture. The working routines of print journalists when using the Internet as a source—in the singular context of Zimbabwean newsrooms—are explored by Hayes Mabweazara in the last chapter of the section.

The second part of the book focuses on the (re)definition of news values and formats on the Web. Jannie Møller Hartley revisits the news categories identified by Gaye Tuchman in her seminal newsroom ethnographies of the 1970s and assesses what has changed in the era of immediacy. Stepping away from the central focus of most online newsroom ethnographies, Steen Steensen focuses on the genre of feature journalism and explores how newsroom discourses and practices of online news have shaped the characteristics of this often overlooked form of online journalism. The section closes with a case study on the coverage of an iconic news event, the 2008 US presidential elections, by Brooke Van Dam.

Section three seeks to invite the reader to move beyond the centrality of the online newsroom as the object of study for Internet news ethnography. We acknowledge the key role of the newsrooms in the decision-making processes that shape innovations, but more and more there are other actors that need to be taken into account to provide a thorough and comprehensive assessment of the evolution of online media. Andy Williams and colleagues show the appropriateness of conducting multi-site observations when dealing with a complex development such as the management of user-generated content in a big public broadcaster. Co-editor Chris Paterson draws attention to the evolution of international news agencies and how their efforts to integrate their newsrooms shape the materials that online journalists work with. Amira Firdaus points out the dire implications of market logic in the configuration of online news services through the case of an Asian national news agency. A final chapter by Chris Anderson provides a research agenda to explore "local news ecosystems," putting the online newsroom in the context of the blogosphere and sources of a specific city, in order to fully understand the dynamics of news production. Pablo Boczkowski offers an Epilogue to the book reflecting on the findings of newsroom ethnography in the context of online journalism studies and the challenges that should be addressed by scholars in the future to keep untangling the evolution of Internet news .

Taken together, the chapters of this second volume analyze over a dozen online newsrooms of all sizes and backgrounds: regional and national online newspapers in several countries; a pioneering net-native like Salon.com; a multimedia regional group with an integrated newsroom producing television, print and online media[5]; national and international news agencies; and three national public broadcasters. We believe that the diversity of cases strengthens the core rationale of the ethnographic approach: technological innovation takes place locally and in each case the solutions adopted may be different. The fact that there are many similarities in the production practices across online newsrooms is better explained if we understand how different settings shape the characteristics of online news.

The geographical diversity of the research presented in this volume is another aspect to highlight. This volume analyzes twelve experiences from seven countries: Belgium, Denmark, Malaysia, Norway, UK, USA and Zimbabwe. The Nordic countries with four cases and the Anglo-Saxon culture with five are clearly overrepresented, but that only reflects the higher quantity of production studies in those areas. Why is there more ethnographic research being conducted in Northern

Europe and the USA? Structural reasons regarding resources for research and the prevalence of journalism studies at academic institutions may be one factor, but also the fact that the development of online journalism has been often led by American and European media may explain a greater scholarly interest in the phenomenon in those countries. Altogether, the two volumes of *Making Online News* cover cases from all the continents, offering a fairly comprehensive perspective on global trends in online journalism.

. . .

We hope this second edition is as inspiring as the first one and that more and more media scholars feel compelled to wear an ethnographer's shoes and spend time among online journalists. We vindicate the value of detail and the explanatory power of narrating the nuances of everyday practices. As Coleman puts it: "This anthropological imperative posits that the devil is in the details; these details are often aesthetically valued for revealing the splendor of sociocultural life and at times are also ethically deployed to push against faulty and narrow presumptions" (2010: 11). The authors of the chapters in this volume observed journalists at work during a timespan of some months to some years, usually in visits scattered over time. Some of the most unique materials that this book provides are narrations of events observed by the researchers; moments in the life of the newsrooms that are revealing of the values, aspirations and constraints of online journalism.

A singular case is the one of BERNAMA, the Malaysian news agency (Chapter 11), where Firdaus explains how access to do ethnography was barred[6] and ended up using only one of the tools of ethnography, in-depth interviews, normally a complement to observations in the rest of the chapters. The analysis of audience participation management at the BBC (Chapter 9) included up to 115 interviews. Møller Hartley, Steensen and Van Dam have in common that their chapters add a valuable triangulation to their observations by doing content analysis of the final product. Some researchers managed to get access to the content management system and the online communication tools of the newsroom, a crucial aspect in the social dynamics at the workplaces that media companies do not always allow observers access to (Puijk, 2008).

A final word for the road ahead. So far, the focus on the newsroom has been very fruitful to demonstrate what makes online journalism what it is. And it will still be crucial in the future; ethnographic data open up the black box, as Latour (2005) would put it, of the network of decisions and definitions made and used by those who produce online news. News websites are what newsrooms have decided they will be, within the constraints of the cultural and organizational settings of their companies. But we also acknowledge Boczkowski's suggestions in the Epilogue for future research avenues: Moving on to comparative studies between media

industries, and reaching out to understand the relationship between news production and consumption. It may be due time, as Anderson very graphically puts (see
Chapter 12), to "blow up the newsroom" and look beyond it–inside the media
company and outside it. On one hand, the whole media company, rather than just
a specific newsroom could be the unit of analysis, as Boczkowski and Ferris (2005)
or Klinenberg (2005) already have proven successfully when investigating convergence experiments. Observing the marketing decision-making processes would be
crucial to assess how they shape journalistic work in a time when business pressures
are fiercer than ever. On the other hand, ethnographers should assess the dynamics
of the relationship between journalists and citizens at news websites and social
networks, combining virtual and in-situ observations both in the newsroom and
following engaged citizens to understand how they produce stories and what moti-
vates them to contribute reporting and commentary. Groundbreaking experiments
of pro-am collaborations and crowdsourcing also deserve attention in order to
evaluate their ability to generate new forms of information gathering and storytelling. We still need to know more about online newsroom practices for original
reporting, which the first volume of the book suggested was rare: when does it
happen and how are resources allocated? The differences between profit-driven
media ventures and the new projects of non-profit journalism could also illustrate
that, in the end, there is still room for many different ways of understanding online
journalism.

ENDNOTES

1 A biennial conference organized by Cardiff University since 2007 is precisely entitled "Future
of Journalism" and sponsoring journals *Journalism Studies* (issues 9:5 and 11:4) and *Journalism
Practice* (2:3 and 4:3) have compiled selected research presented in that meeting. The tenth
anniversary issue of *Journalism* (June 2009, 10:3) was also devoted to this discussion as well
as the spring 2010 issue of *Daedalus* (139:2), the journal of the American Academy of Arts and
Sciences.

2 See, for example, in the USA: ProPublica.org, VoiceofSanDiego.org, TexasTribune.org,
HuffingtonPost.com; in the UK: HelpMeInvestigate.net; in France: Rue89.fr, Mediapart.fr;
in Belgium: Apache.be; in Spain: Bottup.com.

3 See Mitchelstein and Boczkowski (2009) for a review of online journalism research between
2000 and 2008, and Deuze (2008a) for an assessment of the contributions of online newsroom
ethnographies in the first volume of *Making Online News*.

4 See, for example, the report on news consumption habits in the USA (Purcell et al., 2010).

5 This case, Nordjyske Medier, is part of the analysis of both Bechmann's and Møller Hartley's
chapters.

6 See Paterson and Zoelner (2010) for a discussion about access and professional experience in
news ethnography.

PART ONE

Evolving Newsrooms

CHAPTER ONE

Fostering a Culture of Collaboration
Organizational Challenges of Newsroom Innovation

Steve Paulussen, Davy Geens and Kristel Vandenbrande

Innovation in newsrooms is often not a straightforward process; indeed, it frequently faces many hindrances. This chapter offers a comprehensive catalog of the factors involved in the evolution of the organization of labor and practices in the newsroom, from technological infrastructure to professional profiles, training, journalistic culture and the representation of users. The authors provide a review of the findings of recent research on convergence and online journalism and contribute their own ethnographic analysis of two cases of Belgian newsrooms that experimented with collaborative practices: cross-media news production and hyperlocal citizen journalism.

EDITORS' NOTE

Innovation has become a key concept in newspapers' business strategies since the late 20th century. In their review of the literature on organizational development, Gade & Perry (2003, p. 328) refer to research indicating that "companies that value 'innovation' are better placed to use employee creativity to compete in changing environments. Innovative companies are 'integrated' (...) and create opportunities for fresh thinking across organizational boundaries." This description of innovation in terms of 'integration' and 'organizational change' is useful to understand the current discourse of newspaper management. In their search for cost-efficiency and the consolidation of profits, today's newspaper strategies are primarily focused on crossing boundaries and fostering an organizational culture of collaboration and cooperation. The essence of an innovative digital newspaper strategy seems to lie in its attempt to 'break down the walls'—be they between the print and online newsroom, between different outlets of the media organization, between professional journalists and their readers, or even between editorial and commercial departments.

For the purpose of this chapter, two of these areas of 'integration' are of particular interest. The first case that will be described deals with the integration of print and online activities, while the second case focuses on the adoption of user-generated content (UGC) and audience participation in the news production process. In the past few years, the idea(l)s of 'integrated multimedia newsrooms' and 'participatory journalism' have gained a lot of attention within the newspaper industry.

Whereas pioneering examples of newspapers that changed their monomedia workplace into a convergent multimedia newsroom can be traced back to the late 1990s, it took a bit more time for newsroom convergence to attract scholarly attention (Singer, 2004, p. 4). Besides a few normative and somewhat utopian accounts of the emergence of 'multimedia journalism,' the body of academic literature and theory-building on the transition of newsroom organizations is still small and scattered (Gade & Perry, 2003; García Avilés & Carvajal, 2008). Moreover, the rather tentative and cautious conclusions of these academic studies contrast with the many trade reports that express the strong belief of newspaper managers in the necessity, if not inevitability, of newsroom integration. According to the 2008 Newsroom Barometer survey of the World Editors Forum (WEF), for instance, the vast majority of newspaper editors from all over the world agreed that within five years the 'integrated multimedia newsroom' would be the norm for newspapers in their respective countries (WEF, 2008: 7).

Compared to the idea of multimedia reorganization, the growing interest in trends of user-generated content and audience participation in the news making process is even more recent. Again, the opportunities of this innovation are met with great enthusiasm. The trend of citizen journalism has generated a lot of excitement about UGC and citizen involvement in journalism. The idea of participatory journalism is that traditional newsrooms will open the different gates of the news production process for user participation. The final goal of participatory journalism is to change journalism from a top-down lecture into an open conversation between professional journalists and 'the people formerly known as the audience' (Gillmor, 2004).

Recently, academic research gives reasons for newspaper management to lower the often utopian expectations about the adoption of user-generated content in the newsroom. In spite of the progressive adoption of user-generated content by mainstream news organizations (Hermida & Thurman, 2008), empirical studies on participatory journalism show that professional journalists appear to be rather reluctant "to open up most of the news production process to the active involvement of citizens" (Domingo et al., 2008, p. 339). Most journalists try to retain their gatekeeping role and seem to be hesitant to radically change their newsroom routines (Singer et al., 2011).

The discrepancy between how managers see it and how employees perceive it is a recurring factor in the empirical research on the adoption of innovations in newsrooms. Quinn (2005, p. 29) argues that convergence strategies are most likely to be successful if they can satisfy "the twin aims of good journalism and good business practices." While managers are traditionally more occupied with finding new business models, journalists tend to evaluate an innovation on the basis of its (perceived) potential to improve the quality of journalistic work. Gade & Perry (2003) found journalists are more likely to embrace innovations and organizational change if they think they can result in better journalism: "journalists are more open-minded toward change and report higher job satisfaction when change is tied to pursuit of journalistic, not business, goals" (Gade & Perry, 2003, p. 329).

Two Case Studies

Under the subheading 'The Keywords of Convergence,' a contribution in the WEF's Trends in Newsrooms 2008 report stated that "conviction" and "communication" are critical to achieve the newsroom's "cultural change" that was needed (WEF, 2008). For newsroom management, convincing journalists of the potential benefits of innovation and reorganization is indeed an important step to change newsroom culture. But it is only one of the steps. Rather than considering organizational change merely as a matter of 'communication' and 'conviction,' this chapter tries to make clear that the adoption of innovation in newsrooms is a social process that is shaped by the broader context within and outside of the newsroom. As Gade & Perry (2003, p. 329) conclude from their review of organizational development literature, "changing an organization's culture is not easily accomplished, and understanding the need of change is often not enough to convince employees, and even members of management, to accept change."

By taking a social-constructivist approach to the adoption of innovations, we try to get a better understanding in the professional and organizational factors that foster or hinder the acceptance of change in professional newsrooms. Theoretically, the research presented in this chapter is situated in "the third wave of online journalism research," which focuses on the "process of innovation" rather than on "the effects of innovation" (Domingo, 2008c, p. 16). Assuming the mutual shaping of technological and social developments, the focus of these studies, often based on qualitative case analyses, is on the "dynamic relationship between technology, social actors and context factors" (Domingo, 2006a, p. 296).

The empirical data presented in this chapter are based on two diverse newsroom studies, in media companies with dissimilar visions and using slightly different methods, to illustrate some recurring challenges of the processes of bringing a vision into practice through organizational change. Since no two examples of newsroom reorganization are exactly the same, these case studies do not pretend to be representative or comparable, but they both illustrate the complex interactions between (technological) innovations, on the one hand, and the routines, roles, workflows, structures and cultures within a newsroom, on the other hand.

The data of the first case are based on an ethnographic study at the Belgian newspaper company Corelio. A four-month participant observation was done between March and June 2007. During this period their daily newspaper, *Het Nieuwsblad,* started developing a local news website for each of the Flemish municipalities. Local journalists from both the print and the online newsroom were involved in this operation. Two part-time local news supervisors were appointed to coordinate this cross-media project, one of the authors being one of them. He was present in the newsroom three days a week. His active role in the newsroom authorized him to attend staff meetings and follow discussions (e.g., in newsroom lobbies or by reading email traffic). Throughout this concrete period of transition, routines, practices and roles were intensively (re)negotiated. Contradictions and disturbances—theoretical concepts borrowed from labour and organizational sociology—shaping the relationship between the converging professional roles were observed.

The second case study involves the newsrooms of the Belgian newspaper company Concentra Media, which owns two regional daily newspapers, *Het Belang van Limburg* and *Gazet van Antwerpen.* Most of the data presented in this chapter have been gathered within the framework of a research project on the opportunities of participatory journalism practices in the newsrooms of both Concentra newspapers. The goal of this empirical study was to gain insight in the organizational and professional factors that shaped the adoption of user-generated content by professional journalists. Loosely following the analytical model provided by Boczkowski (2004b), questions focused on organizational structures, work practices and routines and journalists' attitudes towards user interactivity. In sum, 20 in-depth interviews with members of the print and online newsrooms on different levels (newsroom management, editorial staff and IT employees) were carried out during April–June 2007 (see also Paulussen & Ugille, 2008).

Leaving out of consideration the concrete application of innovation—it can go from print-online integration (the Corelio case) to professional-amateur collaboration (the Concentra case)—challenges for media companies largely remain the same. Despite their diverse nature, the case studies can teach us something about more common organizational issues possibly hindering or fostering change in other newsrooms. The findings from our two case studies will be used to further illustrate how organizational context factors, and the complex interplay between them, continuously shape and affect the innovation process.

Visions of Newsroom Innovation

Corelio and Concentra are two examples of the increasing number of media companies that decided to restructure their newsrooms to survive in this radically changing and converging media market. Corelio clearly opted for an integrated newsroom strategy aimed at breaking down the walls between formerly separate (medium-specific) newsrooms. Although the online and print newsrooms are located at the same floor, their desks are clustered according to delivery platform. To foster communication between journalists from these different media, key people from the online desk are sitting in the middle of the print desk. The management of the project of setting up a regional news website was delegated to a local news supervisor, who had to act as a bridge between the regional print desk and the online desk. Although being physically located close at hand to the print desk, the supervisor had fewer personal contacts with its staff (a chief-editor for each regional edition, copy editors, designers and journalists) than with the online desk and was often seen as an outsider.

At first glance, the strategy of Concentra is a bit more ambivalent. In 2008, the company decided to split up its print and digital activities into separate 'business units,' which means that print and online editors are now—literally—both physically and financially disconnected from one another. Although one would think that such a newsroom structure is not favourable to a 'culture of collaboration,' Concentra does try to enhance collaborations within the editorial staff and also between professional editors and the public. The media company was among the first in Belgium to set up a hyperlocal citizen journalism site called *HasseltLokaal* and to hire an online editor for the position of community manager in 2007.

What is similar at both newspaper companies is that news workers previously responsible for only one distinct part of the news production process, obtained new assignments and job descriptions. Media managers at Corelio and Concentra encouraged different kinds of multimedia cooperations and collaborations across former boundaries. They shared the belief that convergence of newsgathering, processing and distribution has to result in scale advantages and thus increased cost-efficiency. As a consequence, news production is also increasingly becoming a matter of teamwork. Journalists that traditionally used to work individually now have to collaborate with people across (e.g., with online news editors, engineers, marketing people) and even beyond (e.g., with citizen journalists) the borders of their own familiar news desk.

Through our observations and conversations with the newsroom managers and editorial staff of Corelio and Concentra, we found that almost all people involved felt that restructuring the newsroom requires a change of culture. At the same time, the people involved recognized that such a cultural change is not easily accomplished. Many authors have stressed that cultural change implies "cultural clashes" (Silcock & Keith, 2006, p. 612). Gade & Perry (2003, p. 340) therefore suggest considering newsroom restructuring as a process of ebb and flow, since the complex interplay between various structural, practical and professional factors makes it difficult for change to occur in a linear way.

From Vision to Practice: Organizational Challenges

When looking at the translation of visions about newsroom innovation into practice, literature confirms that "reality is still differing from wishful thinking" (García Avilés et al., 2008, p. 11). The implementation of innovations in the newsroom tends to be slowed down by the newsroom staff who are not always eager to accept change. An explanation often given by media managers as well as scholars is that journalists are conservative and resistant toward technological innovation. However, even though journalists' sceptical and resistant attitudes do play a significant role in the adoption of newsroom innovation, other practical and structural factors are equally important.

On the basis of literature review, supplemented with data from our own two case studies, we will discuss a range of different and intertwined factors that influence the adoption of newsroom convergence. We do not pretend to give an exhaustive list but try to illustrate that due to the diversity and complex interplay of these factors, successful newsroom reorganizations cannot be easily predicted. The factors we identified relate to organizational characteristics of the newsroom (its physical and hierarchical structure and technological infrastructure), the available human resources (job profiles, level of multi-skilling among the workforce, salary and reward systems), the organization of labour within the newsroom (allocation of time and resources, work routines and workflows) and the professional attitudes toward newsroom innovations (business and journalistic goals, representations of users).

Physical Structure

As said, newsroom innovation strategies are quite diverse. Yet, the different case studies described in the literature make clear that traditional newsroom organization tends to promote internal competition rather than cooperation. Therefore, a principal goal of newsroom convergence should be to strengthen teamwork and cooperation across departmental barriers (Meier, 2007, p. 11).

Still, finding the most appropriate physical structure is not an easy task. The Concentra case illustrates how cooperation and teamwork can be hindered by the newsroom's physical structure. We found several examples of the fact that departments in the newsroom worked quite separately. Occasional collaborations across departmental boundaries generally resulted from coincidental informal contacts between individual journalists. Management decisions even seemed to cause more internal competition between the different departments by splitting up the print and online activities into different divisions, called 'business units.' Each 'business unit' operated as a separate division within the news organization and had its own management, budget and staffing. As a result, there was little cooperation between the print and online editorial teams at Concentra. The physical separation of both teams hindered interactions and collaboration across departmental boundaries (see also Paulussen & Ugille, 2008).

Hierarchical Structure

Organizational convergence also implies conflicts with the traditional medium-specific newsroom structure. Formerly, newsroom workers were involved in only one distinct activity, with particular goals. In a multi-platform newsroom, new tasks in their job profiles engage people in different activities and actions, with different goals and sub-goals (e.g., making content for print and online). Accordingly, the hierarchical structures and cultural relationships (Finberg, 2002) in which they operate, as well as the criteria for rewarding and appreciation of their work, become much more complex. "All of a sudden I have two different bosses," said a television journalist interviewed by Singer (2004, p. 14).

An unclear 'line of command' possibly results in apathy towards organizational development, as we could observe at Corelio. The content management system used by the newsroom of *Het Nieuwsblad* assigned content to the right webpage by using the so-called 'tagging information' (e.g., city name and province). These detailed tags had to be provided by the journalist or copy editor. Yet, they were often incomplete or simply forgotten, despite several reminding emails from the deputy editor-in-chief. A large part of the staff continued to use the former, less detailed, way of tagging, which met the needs of the print edition but was insufficient for the online platform. They still perceived their main task as making a perfect printed newspaper, which was also the main priority of their direct chiefs. The website was not seen as their responsibility and online editors had little incentive to change old habits.

Technological Infrastructure

Also at the Concentra newsroom, we found that some journalists tended to retain old habits. When newsroom management implemented a new content management system (CMS), it appeared that some print editors kept using the old editorial software, as it was still running. One of the reasons for their refusal to use the new system was that the new functionalities were seen as too complicated and counterintuitive to the journalists' work routines. This finding is consistent with Domingo's observation that technical tools need to be sensitive to the journalists' needs and routines (Domingo, 2006b). Furthermore, it is important that new technology works without too many flaws and errors.

At Corelio, the majority of local content for the regional news pages originated from the print newspaper. Every night the articles were automatically exported from the print system to the content management system used by the online desk. Each morning around 6 A.M., an online editor had to check the export procedure. During the observation period, this automatic export failed almost every day, and the online editor had to repair the exported articles manually, which created a lot of frustration.

Job Profiles

Today's mainstream media are reshaping news work in fundamental ways. In their attempt to "do more with less staff, budget, and resources," media companies are trying to create "a more flexible, multi-skilled and highly moveable—at least in the eyes of management—workforce" (Deuze, 2007, p. 147). The result is an increase in "atypical" working conditions, as it has been labelled (IFJ/ILO, 2006), and new job descriptions that tend to add coordination, cooperation and management tasks to the traditional functions of newsroom staff.

We already mentioned that Corelio hired a new editor for the supervision of the regional news websites. This bridging and coordinating role did not exist before. Also in the Concentra newsroom, new job profiles have been created. For instance, the online editorial team now has a 'community manager,' who is in charge of moderating user-generated content and managing interaction with readers.

Training and Multiskilling

There is a general consensus that media convergence requires new skills. Research also reveals a general "perception among journalists that they received inadequate training for work in a different medium, if indeed they received any training at all" (Singer, 2004, p. 11). Other studies confirm that despite the attention paid to journalism education and multiskilling in scholarly literature, journalists still find they receive insufficient or inadequate practical training (Huang et al., 2006).

According to the IT employees at Concentra, one reason for the resistance of some professional journalists to the new content management system (see above) was their lack of basic skills to work with new technologies. This complicated the in-house training sessions. The journalists, for their part, claimed that their technical knowledge was good enough but said that the organized training sessions were too short, and the shift to the new system was too abrupt for them to acquire the skills for using it efficiently.

Salary/Reward

Not only training but also rewarding journalists seems to be critical for successful newsroom reorganization. Singer (2004, p. 12) observed that journalists are well aware whether colleagues are rewarded for contributions to convergence, adding that "for the most part, they are not."

Some of the freelance local correspondents at Corelio considered their personal investment (in terms of material and time) not in balance with the reward. Contrary to the correspondents in employment (with a fixed salary), the freelancers at Corelio were remunerated per blog entry. Due to the relatively small amount of money paid per entry, it was fairly difficult for the local news supervisor to motivate them to update their blog regularly. A vast number of freelance correspondents deemed they could better invest their time in writing better-rewarded articles for the printed newspaper. Per picture printed in the paper, the photographer earned two and a half times more than the remuneration of the correspondent for one blog entry. And moreover, for this, the correspondent was expected to write text, take pictures and to create 'some interactivity.'

Time and Workload

The idea of the multiskilled journalist, responsible for newsgathering as well for capturing sound and images and doing the news processing, is often unpopular among news workers. Singer found that a majority of professionals had no problem with new skill building, but that openness to the idea of convergence was strongly related to perceived workload. "(R)eporters felt strongly that they were being asked to do a lot more work for little or no extra pay" (2004, p. 12).

The two case studies confirm that a lack of time caused online journalists to refrain from taking up new tasks, even when they believed these tasks were gaining in importance. The journalists expressed their concerns about available time, resources and manpower. As a result, both at Corelio and Concentra, journalists said that even though they were prepared to increase collaborations within the newsroom, the workload and time pressures did not allow them to do anything that went beyond their core tasks.

Routines and workflow

Different workflows and work rhythms can slow down cross-media cooperation as well. While the online desk has to handle a 24-hour news cycle, the print desk used to work with one fixed deadline every evening. Tensions can arise between routines and work practices of previously distinct professional groups (not only journalists) within the media company (Huang et al., 2006).

Initially, there was no support for the online newsroom from Corelio's technical staff before 7 A.M., because its working hours were attuned to the work rhythm of the print product. At the urgent request of the online chief-editor, the helpdesk finally assured technical support in the morning as well.

Business and journalistic goals

As mentioned before, Gade & Perry (2003) found journalists tend to be more resistant to innovation if they think it only serves business goals. Even if journalists are strongly convinced of the economic necessity of convergence, it is important to stress the potential journalistic benefits of newsroom change. Quinn (2005, p. 37) therefore says that "under wise leadership, convergence offers opportunities to do better and more socially useful journalism."

Yet, not all journalists are convinced about the journalistic benefits of newsroom innovations. Although most journalists at Corelio and Concentra were positive toward the ideas of multimedia and interactivity, our findings suggest that many of them were concerned about the impact of newsroom convergence on journalistic quality. Much has been written about the negative implications multimedia journalism may have on the quality of journalistic work. The fear of becoming a 'jack of all trades, but master of none' is still observable among parts of the newsrooms. Also studies on participatory journalism suggest that journalists' professional quality concerns constrain professional-amateur collaborations in news making (Thurman, 2008; Singer et al., 2011).

Our interviews with journalists at Concentra confirm that they struggled with the credibility, value and relevance of user-generated content. Especially the often (hyper)local focus and personal tone of user contributions made some journalists wonder whether adopting user participation could result in better journalism.

Representations of Users

Boczkowski (2004b) considers journalists' "representations of users" as one of the main production factors shaping the adoption process of innovations in the newsroom. He found that the use of interactivity by online journalists partly depends on whether they see their users as consumers or producers of content. We found support for the idea that journalists' perceptions of their audience play an important role in the adoption of participatory journalism (cf. Paulussen & Ugille, 2008).

The Concentra newspaper *Het Belang van Limburg* has a strong community commitment and therefore tries to pay much attention to reader contacts. Readers are encouraged to submit pictures and personal stories destined for special human-interest sections in the newspaper or on the website. Reader input is thus primarily associated with '*faits divers*' or 'soft news' rather than with supposed hard news. In general, we found journalists from the regional news desk and the online news desk to be more positive toward user-generated content than editors from other departments.

Conclusion

More and more media companies are making the transition from mono-media to cross-media workplaces. Translating vision into practice is a major challenge for today's media leadership, since there seems to be a discrepancy between 'how managers see it' and 'how employees perceive it.' Implementing innovations and fostering a culture of collaboration in the newsroom tend to be slowed by newsroom staff who are not always eager to accept change.

We have tried to explain that even though journalists' sceptical and resistant attitudes do play a significant role in the adoption of newsroom innovation, other practical and structural factors are equally important. On the basis of literature review, supplemented with data from our own two case studies, we discussed a range of different and intertwined factors that influence the adoption of newsroom convergence, such as training, rewarding, hierarchical and physical structure, routines or workflow. We tried to illustrate that due to the diversity and complex interplay of factors, successful newsroom reorganizations cannot be easily predicted.

ACKNOWLEDGMENTS

This chapter is primarily based on research in the framework of a multidisciplinary strategic research project on Flemish E-publishing Trends (FLEET), which was funded by the Flemish Institute for Science and Technology (2006–2010). The ethnographic case study of Corelio has been carried out within this project. The case study of Concentra is part of a two-year research project (2007–2008) called CoCoMedia (Collaborative Community Media), funded by the Flemish Interdisciplinary Institute for Broadband Technology (IBBT). An earlier version of the chapter was presented to the 2009 Conference of the European Media Management Education Association (EMMA) in Paris.

CHAPTER TWO

Closer Apart? The Networks of Cross-Media News Production

Anja Bechmann

This chapter challenges the idea of physical proximity as the driving force in the structuration of converged newsrooms. Using the case of a cross-media initiative in Northern Denmark, the author discusses how the analysis of networks of interests depicts a clearer picture of the rationales behind creating a "superdesk" to coordinate the journalism of different media platforms. The centrality of the newspaper division is clear as well as the marginalized role of actors inside (photographers, local news reporters) and outside the newsroom. The study suggests that decentralized strategies in organizing the newsroom, rejecting space as the definition of convergence, would allow for a more inclusive approach to active publics.

EDITORS' NOTE

One consequence of converging news production is the integration of the formerly distinct journalists and work processes from newspaper, TV, radio and digital platforms into shared spatial arrangements. For some companies this means moving into shared buildings, while others have established so-called superdesks or multimedia hubs. Superdesks are spatial workplace arrangements where editors from different media are placed next to each other to enhance knowledge sharing and coordination across media (see also Huang et al., 2006; Singer, 2004; Bødker & Petersen, 2007; Erdal, 2008; Hemmingway, 2008). As described in most articles on convergent newsrooms the idea was and is that the closer the different media become, the more important it is for the work of journalists from all kinds of media to be integrated, and the better the cross-media production should be. This idea builds on the assumption that strong cross-media news production is based on spatial proximity (Kraut et al., 2002), but what are the consequences of this idea and to what extent do they achieve the desired result?

The aim of this chapter is to analyze and discuss *spatial proximity* as a core rationale of superdesks, which aims to enhance cross-media collaboration and reach what Singer calls the "the final stage of diffusion, at which the innovation [convergence] has been fully incorporated into the routine activities of the organization" (Singer, 2004, p. 15). This chapter intends to revisit these assumptions from a dynamic and less techno-deterministic approach.

Most literature on convergent newsrooms has discussed the ways in which the shared buildings and superdesks have not fully integrated the work routines, mindsets and identities of the journalists (e.g., Singer, 2004; Meier, 2007; Deuze, 2008b). This chapter contributes in two ways to this discussion: First, it gives an in-depth snapshot study of collaboration patterns at the superdesk of a Danish integrated newsroom and discusses the implications of spatial proximity by using the analytical concept of networks. Second, by drawing on interviews and observations conducted over a period of five years it suggests cross-media news production may be better understood if, instead of analyzing it as a linear process or through outlining its constraints, we consider how different groups of interest negotiate what and who is included and excluded in the cross-media production process at any given time. The chapter will focus on analyzing the dynamics of a superdesk as a case of spatial proximity and how these dynamics influence remote news production participants such as local news departments and users.

The Concept of Networks and Spatial Proximity

The chapter will make use of *spatial proximity* and *networks of interests* as two central analytical concepts. The concept of networks in this chapter is closely connected to the work of Castells (1996, 2009) and Latour (1986, 1987). Castells suggests that contemporary societies should be understood as a collection of networks, where the operating units are interactions initiated by shared or negotiated interests rather than institutions (Castells, 1996, p. 187). Following this line of argument, there is no inside and outside of an organization, only inside and outside of networks, along with the closer and weaker ties they instantiate. Therefore, it is relevant to discuss whether a spatial centralization of interaction *within* a (networked) organization contains (or does not) all the relevant parties and interests in the network. In other words, the journalists sitting in the spatial arrangement, the interests they represent, and the interpretations and interests that are left out may define the borders and limits of the news production network. This approach, therefore, can offer a more realistic assessment of the implications of centralized superdesks in the production workflows of convergent newsrooms.

What defines the network is negotiated between different actors (Latour, 1986; Castells, 2009). Spatial arrangements such as the superdesk can be used as a way to define the network. Those that define the network according to Castells are called *programmers*, the ones that decide the exclusion and inclusion rules, setting the standards for the network, constituting the networks and thereby "programming" the network. *Switchers* on the other hand are those that combine and intertwine different networks through shared goals and resources (Castells, 2009, pp. 42–47).[1] In the context of cross-media news production programmers would be the ones that decide on what is (good) cross-media production and, for example, what should be the priority and pace of the production, whereas switchers would be the ones that negotiate different media and production interests.

Taking a Latour-inspired point of view, spatial proximity in the arrangement of a workplace such as the superdesk could be an *actor* that both defines and delimits the network.[2] On the other hand, *place* plays another fundamental role for Castells because he assumes that there is a thing called network society where *space of flows* is a constituting characteristic. What characterizes the space of flows is a mixture of (physical) places, material communication networks, and the content and geometry of the flows of information performed by social activities (Castells, 2009, p. 34). Spatial arrangements such as the superdesk are heavily connected to other "places" through different media platforms such as telephones, digital production systems, email and chat systems, and community Web pages (for the news consumers and stringers) making the physical ties less important. On the other hand, establishing the superdesk as a physical control unit may remove the control from the decentralized units in the network such as local departments and the users.

Why then would media companies use spatial proximity as a guiding principle for cross-media when newsrooms are interacting with networks in and out of the organisation both in terms of research, content production and distribution channels? Turning to studies on distributed versus non-distributed collaboration among scientists, Kraut et al. (2002) suggested that physical proximity means that it takes little effort to start interacting and there will be repeated encounters and frequent communication. This increases the likelihood of collaboration (Kraut et al., 2002, pp. 140–142). Also, it is easier to communicate because the nonverbal side of communication (over 90% of the total communication according to Meyrowitz, 1985) such as pointing is more easily accessible. Moreover, passively monitoring activities (picking up relevant information without directly interacting), and the awareness of sharing an environment and a team are an essential part of why spatial proximity is understood as a success (Kraut et al., 2002, pp. 154–156). However, authors also suggest that collaboration in co-located settings seems to be over-attentive to 'local' accessible information in the working groups at the expense of the more remote.

Inspired by urban design and architectural theories, Harrison & Dourish state in their study of computer supported collaborative work that the "focus on spatial models [for collaboration] is misplaced" (Harrison & Dourish, 1996, p. 1). Following Castells, they suggest instead to distinguish between *space* and *place:* "space is the opportunity, place is the understood reality" (ibid.). Referring to Goffmann's (1959) idea of acting on different stages, Harrison & Dourish suggest that collaboration, as an act, is determined as much by the presence of certain individuals or groups as by the location itself.

If we use this theory within converged newsrooms we could suggest that physical proximity plays a minor role compared to groups of interest. A place can then both be seen as a part of, and apart from, a space (or spatial proximity), depending on how the distinctiveness of the place is defined. If defined by the physical proximity, participants may then as suggested by Kraut and colleagues (2002) be less reluctant to create places with remote participants. If defined by interests as suggested by Latour and Castells, the use of newsroom designs that enhance proximity could be considered counter-intuitive.

Introducing the concept of *networks of interests*, which function as shared "places" across physical space, adds a new dimension when examining what kind of places are created in the converged newsroom. Are production workflows spatially bound by physical proximity or are they organized in terms of interests? In this chapter I present findings from my ethnographic study in a convergent newsroom, using the concepts of a non-physically bound *place* and *networks of interests* as analytical tools to reinterpret spatial proximity.

A Case of Cross-Media News Production

This chapter is based on an ethnographic study of cross-media production routines in a Danish news organization. Nordjyske Medier is a media corporation in Northern Denmark producing news for five different media platforms. The news corporation was a pioneer in cross-media production in Europe when they introduced their integrated newsroom in September 2003 (Northrop, 2005) and is still one of the leading cross-media news players in Denmark.[3]

The study was conducted over a period of four years from 2004 to 2009 to follow the development of cross-media production over time and within different news genres. In 2004 and 2005 I, together with three colleagues from computer science, observed meetings (September 2004) and the activities of new roles such as media editors and cross-media coordinator, as a part of a multi-industry research project on new work practices. The cross-media coordinator is the chief editor in the newsroom, whose job is to decide the priority of the stories and which stories should go to what media. We also made snapshot observations of the different departments within the organisation to understand collaboration and production routines.

In this case, the physical space seemed to play an important role in cross-media production. The understanding of a converged newsroom by the management literally meant moving journalists into the same building with shared desks, meeting facilities, and coordinating arrangements such as the central cross-media editing desk which they called the "superdesk" (see Figure 1). All journalists were encouraged to produce for all platforms, but the media editors were responsible for only one platform (newspapers, digital platforms, TV or radio). Placing the editors physically next to each other would enhance cross-media news production.

FIG. 1: The superdesk: A news editorial work place and meeting facility that builds on the thought of spatial proximity as a strong tool for collaboration

The focus on spatial proximity led to experiments with different space-syntax methods (Steen et al., 2005) in March 2005, when we tried to register and code the interaction between the journalists on the superdesk. This gave 75 snapshots that were used to identify who was interacting with whom. It gave an insight into the frequency of verbal interactions, but it was complex to code non-verbal and mediated interactions (e.g., emails and chat). We tried to observe how information technology was used as a tool for cross-media collaboration, but it was difficult to observe it in a systematic way. The space-syntax methods used in this chapter, therefore, under-represent online interactions and passive awareness of the other media as suggested by Kraut et al. (2002). However, we asked the journalists about the passive awareness in the interviews afterwards.

Overall, we were four people spending approximately fourteen full days of observation over the designated time period. In addition to observation, we analysed documents and conducted 35 semi-structured interviews with journalists (26), administrative staff (2), and managers (7). We also distributed a questionnaire among those journalists we could not interview. The interviews focused on the transition to cross-media news production, spatial arrangements, cross-media versus single media workflows, professional skills, and journalistic identity. Follow-up research continued until 2009.[4]

The methodological description shows that the case study was designed as an in-depth study of a pioneer case and can therefore only serve as an illustrative example of the interaction occurring on a superdesk and in a converged newsroom at a certain time. What is special about this methodological design is the emphasis on space coding and the combination of systematic observational registration with qualitative interviews and studies of different production routines across news genres. This gives a thorough insight into the different cross-media interpretations and the different (spatial) cross-media challenges in the networks of news production.

Spatial Proximity and Cross-Media Collaboration

The interviews in my study show that spatial proximity between different media editors and journalists is considered one of the most important things when it comes to cross-media news production at Nordjyske Medier. For the journalists, the benefit is, for example, learning something from colleagues; and for the management it has a corporate value because the journalists, according to the managers, are more inclined to share identity on a corporate level rather than a medium-specific level. This idea is consistent with the findings of similar case studies. Singer (2004) concluded "the trust-building benefits of physical proximity seem to extend outside the newsroom and into the corporate realm" (Singer, 2004, p. 15).

The superdesk was mentioned most when interviewees were asked about spatial arrangements, so reinforcing the notion that spatial proximity is of importance when it comes to the *understanding* of cross-media collaboration and cross-media news production. However, to go deeper into the meaning of spatial proximity we have to ask what kind of collaborations are supported by spatial proximity and how this may affect the managers' and journalists' interpretation of cross-media news production.

The activities on the superdesk consisted of formal meetings, informal meetings (ad hoc), and some talking across the superdesk; media editors with work stations on the superdesk rushing in and out, running, rolling chairs; journalists collecting printouts of production documents, walking by and through the superdesk, picking up other companies' newspapers, reading newspapers, having phone conversations, watching TV news on flat screens above the desk. Looking at the observations and interviews, the superdesk comes across as a production and work facility for media editors and the cross-media coordinator, a place for daily and weekly editorial meetings, and a center for coordination (see also Bødker & Petersen, 2007). As a work facility, the media editors stated that spatial proximity was very important for communication and knowledge sharing:

> Physical proximity is important. Those, who are placed marginally, could just as well be on a foreign planet. Here [at the superdesk] there is a sense of spatial community. In this way the superdesk works as intended. Situations arise when you see each other (Media editor, 2005)
>
> You answer the phone and instantly colleagues hear it (...) it is important to know, what is going on (Project manager working at the superdesk, 2005)

The two statements support Kraut and colleagues' (2002) theory of physical proximity enhancing passive awareness, but it also seems to suggest that there is a space-bound place established at the superdesk with the media editors as participants. The observation of the work routines at the superdesk showed frequent use of informal meetings consisting of short or long work-related conversations, mostly for coordination between media editors. The coordination was often about the priority of the media platforms, which medium should break the story first and how should the perspectives of the media stories differ. Negotiation of the allocation of scarce journalistic resources on the other hand often took place at the editorial meetings, one in the morning and another early in the afternoon.

However, the study also showed uneven verbal interaction patterns among the different media editors. Despite the physical proximity of all editors, some editors interacted (verbally) more than others. When we counted interactions on the superdesk from our coding schemes, the following patterns occurred (Table 1 on the next page).

Those who interacted most were the newspaper and Web editors; the TV editor and the media coordinator; the newspaper editor and the media coordinator; and the TV and radio editors. Those who interacted the least were Web and photo editors, newspaper and radio editors, Web and TV editors, the Web editor and the cross-media coordinator, and the radio and photo editors.

TABLE I: Number of verbal interactions with work-related content among editors at the superdesk. The coding was made every five minutes on two days.

	Newspaper	Web	Radio	TV	Photo
Cross-media coordinator	18	2	12	19	4
Newspaper		25	1	7	2
Web			14	2	0
Radio				17	2
TV					6

To understand *why* these interaction patterns are so uneven despite the shared spatial proximity we have to look at the observations of the different activities on the superdesk and the daily routines of the editors in more detail. Most interactions in the registration are grouped around the newspaper and TV editors, suggesting they are hubs in the news production process. This is also reflected in the editorial meetings where stories for these platforms are the focal point. However, the newspaper editor is not only a powerful *programmer* (Castells, 2009)—we will discuss that later. Simply, the other editors accept the division of power. The observations show, for example, that the Web editor does not participate in the morning editorial meetings because it is about giving birth to and developing stories and "the Web editor does not produce," in his own words. Therefore, he does not see the meetings as an opportunity to negotiate the way in which the stories are produced. In fact his action, by omission, enhances the newspaper editor's interpretation of cross-media news production.

In spite of this power relation, the registration of interaction patterns shows a high degree of interaction between newspaper and Web editor. The reason for that could be the fact that they are sitting just next to one another on the superdesk (see Figure 2), but the observations also suggest that there is a shared network of interests.

Nordjyske primarily interpreted their Web activities in textual and static visual formats which are easily shared with the newspaper. The observations show that the interaction between newspaper and Web often was an informal briefing of the newspaper editor on incoming news agency stories like "have you seen the new story on Joachim?" or coordination such as "should I make a quickpoll for the election?" The high level of interaction registered above primarily took place during the evening, when the website news was cloned from the newspaper.

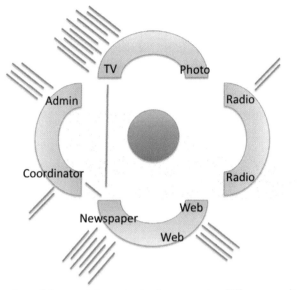

FIG. 2: Illustration of the spatial proximity between the different media editors in the superdesk.

TV and radio editors do not sit next to one another on the superdesk, but the observations once again point to a high level of interaction, which again suggests a shared network of interests. Radio was 'shoveling' sound from TV. Furthermore, the two media editors shared some of the same production routines such as copy writing, editing sound, and creating flow and continuity. The interaction that took place such as "I am printing the TV copy" or "we need a guest for TV" showed these shared interests.

Despite the function of the superdesk as a spatially determined community or place among the media editors, the analysis of the uneven interaction patterns shows that some people interact more than others. This could not only be due to physical proximity, as then they would interact evenly, but rather due to shared networks of interests such as media similarities, similar work routines and shared raw material and news products.

Marginal Proximity and Participatory Journalism

As the superdesk is designed to create a place based on physical proximity that enhances passive awareness, it can be also understood as a network of interest that may exclude people in the newsroom as well as outside the newsroom.

During the observation, a TV director expressed that she was happy with her work place next to the superdesk because the TV people (TV editor, TV director, continuity, and desk reporters) were all sitting close to one another. Consistent with the earlier theoretical discussion this statement shows that she feels part of the network, she is oriented towards her network of interest, and it shows that she finds the spatial proximity strengthens this interest-based network.

In contrast to the TV director, the free newspaper editor who also sits in the same room as the superdesk but in a distant corner feels excluded from the cross-media network of news: "The cross-media coordinator forgets us, except when he talks about audience ratings" (media editor outside the superdesk, 2005). Nordjyske Medier has other local departments spread across the region, thereby being marginal to the superdesk from a spatial proximity perspective. The same feeling of exclusion is present here:

> We have occasionally left messages about stories relevant for TV, but they always get back to us too late. It is irritating that you cannot be a part of the process...sometimes there is a lack of respect for what we do. The local sections are never mentioned in the evaluations. There is nothing about the local news departments and then we stop reading them. (Journalist employed in a local news department outside the building where the superdesk is located, 2005)

Overall the study shows that the people who are placed far away from the superdesk feel more isolated than the ones closer to it. In some cases journalists are content about this because it means 'less work.' In other cases they want to be closer to the superdesk and become more central to the production process as these observation notes from conversation with a photographer show:

> The photographers were placed in the department of sports. It was very isolated according to them. Then they moved so that they were close to the newspaper editor, which enhanced the collaboration according to a photographer. Now the superdesk has come between the photography and the newspaper editors. This means a more structured and official contact with the newspaper but less individual contact (notes from a conversation during observations with a photographer, 2005)

However, the main focus of the interviewees was not proximity with the superdesk as such but with certain important *persons* and *roles* on it. Being closer to them was perceived by journalists as enhancing the likeliness of being seen and heard through passive awareness. In this case, the newspaper editor seems to function as a powerful *programmer* of the cross-media news network because the newspaper succeeded in dictating the time of, for example, the editorial meeting (not suiting the TV routine), the first right to exclusive news, and the morning newspaper being the all-important focus in the evaluation process. The cross-media coordinator as one of the other important persons serves as the *switcher* (Castells, 2009), trying to integrate different media-specific interests with shared goals and standards.

The analysis in this chapter, however, shows that it is not a given that physical proximity necessarily leads to shared interests—and thereby inclusion in the network of news. As in several other studies (Singer, 2004; Erdal, 2008) a certain degree of exclusion has proved to be the case for the Web journalists (see also Bechmann Petersen, 2006).

Outside the organisation, Nordjyske Medier—like many other news producers- has tried to integrate users as a part of the production process, a form of produsage (Bruns, 2008) that ideally makes news production a highly distributed process. Nordjyske Medier has always used stringers as local reporters, but in 2006 they created an online community where users could create news stories, and some of the best stories would then be selected for the free printed newspaper until this free product was closed in 2008. Stories from the newspaper and radio were repurposed to the community to start a discussion among the users, but the journalists did not engage in a conversation about the story. Generally, the journalists paid little attention to the contributions from the users and the news that was created in the community. They thought citizen contributors did not have the standards and skills required for professional news production:

> From traditional news criteria and an understanding of journalism this is not news, but people write about what interests them: "I have this recipe, but the dough doesn't seem to work, can someone help me with some ideas to improve it." That isn't news...then it must at least be within certain topics, where people feel comfortable (new media manager, 2008)
>
> It is not different from sitting at a railway station and listening to other people talk. In one out of 100 conversations there are interesting things being said and then it is the role of the media corporation to find that single one and make it reach a bigger audience (new media manager, 2008)

In the framework of physical proximity and passive awareness as a way to create a shared *place*, the representation of the users in the newsroom is weak. There was a young community moderator sitting in the adjacent room to the superdesk, but she had little interaction with the people at the superdesk during the observations and did not participate in the coordinating editorial meetings. Even when looking at the superdesk as a network of interests where the newspaper editor and cross-media coordinator play central roles (as powerful programmer and switcher), the representation of users is absent.

Decentralizing the news production process through user participation seemed to clash with the concept of converged newsrooms as being about physical proximity and with the understanding of converged newsrooms with a focus on the integration of the mindsets of different *media* editors. One could argue that in this layout the Web editor would represent user interests. However, the observations showed that even though the Web journalists were expected to foster user participation this was not always the case because they had other interests as well. When they discussed the general attitude in the newsroom towards user-generated content, they (being the underdog) agreed on shared understandings of journalistic quality that marginalized the users. Bruns (2008), among others, suggests that empowering the users in the news production process is about letting the users make some decisions (acting as co-programmers), allowing the production standards to be defined within the community network itself and not from a central controlling unit and a hierarchical power structure. This seems to be the direct opposite of a spatial construction such as the superdesk. From this perspective, the superdesk reinforces traditional media hierarchy and acts as an impediment to the quality of content produced by the non-hierarchical user networks. As Hermida and Thurman (2008) have noted, the need to filter, aggregate, moderate and control is closely connected to the journalistic identity, while in the social-networking, user-centred model these functions are to a high degree done by the users themselves.

Physical Proximity, Networks of Interests, and Understandings of Cross-Media

In this case study, the superdesk seems to be efficient when cross-media is interpreted as optimizing "the unexpected" (Tuchman, 1973) for different media platforms with scarce resources. It is also efficient when cross-media is interpreted within the media platforms *owned* by the media company (Nordjyske Medier) even though it favours what the newspaper editor and cross-media coordinator (as one of the managing editors) think is the right priority. However, this understanding of cross-media assumes that the company is in control of the news and the distribution channels and that news is about speed—or what Domingo calls "immediacy" (2008a). Singer (2008a) and Bruns (2008), among others, reject the idea that news is only about speed because there will always be some users or another company that are faster. On the other hand Burgess & Green (2009) have showed that content from established media companies plays a central role in social networking sites such as YouTube.

To focus on the companies' own distribution channels for the news is also fraught with danger. Cross-media storytelling and "Create Once, Play Everywhere" strategies in the controlled channels could now be supplemented with understandings of cross-media as what I have called "Discover and Link," where users find something of interest and send it around on the network (such as linking to the story on Facebook or retweeting), "Fetch and Distribute," where users *steal* the content away from the company and place it on other companies' webpages (such as YouTube), or "Open and Play" where the users contribute with content (or code) to the media universe (Jenkins, 2006) of the company (Bechmann, 2009, p. 138).

Looking at these understandings of cross-media, they seem to challenge the reductionist approach of news companies, because they are interest based rather than media based. Physical proximity loses its central position on defining cross-media initiatives if we look at it from the user perspective.

The study at Nordjyske Medier shows that, over time, certain networks of interest have shaped the way in which cross-media is understood and practiced. The most significant networks of interests, which could be drawn from my study, are built around the ideas of *efficiency, innovation, profession and career*. There are interests in wanting to reduce costs by creating a more efficient news production process (especially materialized with the establishment of the superdesk); interests to include new ways of producing, understanding and communicating; interests in preserving the role and skills of journalists, and interests in using cross-media as a career booster by improving the skills of production across all platforms. The survival of the newspaper industry has served as a shared goal of these different networks of interests.

These, however, do not seem to be interests that can unite actors outside the company, such as Twitter, Facebook and Google/YouTube as main distribution channels or hubs in the network, and the users as target groups. Therefore, it is relevant to discuss whether Nordjyske Medier as well as other news companies may need to *reprogram* the networks of cross-media news production to encompass a shared 'place' with the users. As users' attention and interest are the primary source to build revenue for commercial news companies, a negotiation of how users can be integrated into the production process is crucial. It would be interesting for future ethnographic research to analyze the interests of users in news production more thoroughly.

Conclusion

This chapter intended to find out whether a newsroom constructed around the spatial proximity of the media editors enhanced cross-media production. The findings show that 'place' is strong in the sense of passive awareness, but the study also shows that networks of interests seem to control the interactions occurring on the superdesk and across the physical space in the whole organization.

Even though the superdesk works well in terms of creating a shared 'place' for the media editors sitting in the spatial arrangement, this 'place' can be questioned in terms of what kind of understanding of cross-media supports for the company. The analysis shows that by choosing to put *separated* media editors next to each other, the company seemed to support an understanding of cross-media that builds on the assumptions of control of the news channels and optimizing distinct media distribution channels. This creates a disadvantage when trying to integrate, for example, users in the production process, and visibility in viral networks such as Facebook, Twitter, and YouTube. In this sense, physical proximity is inflexible and preservative unless the constellation of persons and thereby the interests represented are changed at the same pace as the strategies of the media company. In this case study, the superdesk supported a fixed understanding of cross-media because the interests represented at the central place remained the same.

The understandings of cross-media are constructed and change according to the interests of those defining the network. Would it then be more fruitful in a shifting (networked) media landscape and in the cross-media news production to leave the idea of physical proximity behind and bring the production processes closer together by actually being apart in a more distributed structure? The study has shown that it is not the physical space that is the barrier. What defines the production process network is access to certain central persons who have the power to set the inclusion and exclusion rules, that decide who is marginal and who is central, and that combine different actors in shared goals. Looking at the question from the outsider's (local news departments and users) point of view, their access to the central decision-makers might improve if the spatial arrangement is removed. But, our observations point to the conclusion that if the news company wanted to integrate a more distributed news production process (with the users), it is the centralization of control that should be rethought. The spatial arrangement and the focus on physical proximity are just symbols of this. Looking ahead under the light of this study, physical proximity could be displaced as the core aim of newsroom convergence, and instead more dynamic approaches could be used to support strategically fruitful and *shifting* shared 'places,' so that different interests could be represented in the 'control unit.'

ACKNOWLEDGMENTS

The author gratefully thanks DEKAR (project under the Nordic Innovation Center) and fellow researchers Stinne Aaløkke, Susanne Bødker and Eva Bjerrum for co-collecting data for this chapter and Charles Ess for proofreading.

ENDNOTES

1 Castells and Latour do theoretically disagree on certain points, especially when it comes to conceptions of society and power. Also, Castells does not use the term actor for non-human agents, like Latour does. My conception of network relies in the end on Latour's theories, incorporating Castells' concepts to that core.

2 Here Latour represents ANT (Actor-Network Theory) where an actor can be both human and non-human. Strategy papers and spatial workshop arrangements can play an essential

role (as actors) in the negotiation of shared interests and goals and thereby in establishing or strengthening certain networks.

3 Nordjyske Medier is fully owned by the Danish company Nordjyske Holding. The five different media platforms that Nordjyske Medier produces for are newspapers (morning newspaper, free newspaper and local newspapers), a 24 hour news channel, websites, mobile phone and Tele-text. Across platforms Nordjyske Medier daily reaches over 90% of the total population in the Northern area of the Jutland peninsula, in mainland Denmark.

4 In 2006, I made three thematic case studies on my own using similar methods. I studied the production of three different news categories in terms of planning processes (Tuchman, 1973), a premier league soccer match (sports), a local news story, and a news event engaging the community (a weight loss contest with a duration of half a year). In 2007, 2008, and 2009 I had meetings with the head of the new media department where we discussed my findings and the ongoing development within the company. In 2009, I conducted a semi-structured follow-up interview with this manager.

CHAPTER THREE

"Beaming up" Traditional Journalists

The Transition of an American Newspaper into Cyberspace

Sue Robinson

This ethnography documents one of the most radical experiments in recent journalism history: the decision of a US newspaper to stop printing and focus on online publishing in order to survive the economic crisis. The discourses of the management embracing the opportunities of interactivity and multimedia contrast with the attitudes of journalists, confused in the process of redefining their identities and "news" in the new context. The tensions between the traditional values of journalism and the expectations of Jenkins' convergence culture become apparent in the process: the newsroom tried to retain the authoritative voice of watchdog for the community, while at the same time reaching out to satisfy the new imperatives for a successful news product: volume, freshness, connection and experience.

EDITORS' NOTE

In 2008, *The Capital Times* of Madison, WI, elected to end the newspaper it had been publishing for 98 years and move the main thrust of its journalism online with the exception of a twice-weekly magazine. Calling itself a "pioneer" for the industry, the news organization was the first US daily newspaper to do this. By making this decision, the newspaper was hoping to eliminate much of its overhead, recoup losses from a circulation that had tumbled to less than 17,000 from a high of 45,000 and preserve a century-old socially progressive tradition to "comfort the afflicted and afflict the comfortable."[1] In news articles announcing the transition, their headlines declared: "Beam us up!" but the editors and their staff soon learned they had their work cut out for them. How do you maintain entrenched print-world norms while also catering to the digital cultural expectations of their audiences?

In the print world, the mission of *The Capital Times* was executed according to Kovach and Rosenstiel's (2007) *Elements of Journalism*; news should be truthful, independent, dedicated to the citizens, empowering, engaged, relevant, balanced, proportional, and significant, among other things. *The Capital Times* assumed this sense of purpose, even establishing a foundation that funded community-centric projects in Madison, Wisconsin, while declaring itself "Wisconsin's Progressive Voice" on its masthead. But going digital meant a whole new set of standards, some set by technology and not ideology—or at least that is how the editors and reporters sometimes felt. Everything had to be rethought, newsroom leaders realized. Using research from a yearlong ethnography of the newsroom, this chapter will document some of the challenges that had to be overcome in that transitioning organization.

The major question guiding this research asked how a newsroom could transfer the community institution it had built over a century as a watchdog and storyteller into the "convergence culture," a term coined by MIT scholar Henry Jenkins.[2] Two unique conditions characterized this developing culture: interactivity and multimedia. By 2008, many scholars had begun to document the situational differences between media industries marked by digital information's new mobility, omnipresence, and multi-authored surplus. Jenkins (2006) posited that a "convergence culture" went beyond mere combining of mediated formats such as broadcast and print. Instead, the convergence culture represented a philosophical shift merging the old mediated world—and the roles, relationships and rules established therein—with a new, digital one where media are not the only producers of content nor its sole filters, aggregators and interpreters. Now people could be bombarded with every idea, brand, trend, issue and story in multiple domains, public and private, across borders and different media systems—and have agency over that information as well. This paradigmatic change created a new communicative mode—called "mass self communication" by sociologist Manuel Castells (2009):

> It is mass communication because it can potentially reach a global audience, as in the posting of a video on YouTube, a blog with RSS links to a number of web sources, or a message to a massive e-mail list. At the same time, it is self communication because the production of the message is self-generated, the definition of the potential receiver(s) is self-directed, and the retrieval of specific messages or content from the World Wide Web and electronic networks is self-selected (55).

Mass self communication is meant to be considered in contrast to the mass communication (of newspapers, for example) and interpersonal communicative patterns. Castells posited that digital technologies set up new dynamics within the new "network society" that restructured the information power hierarchies. But other implications of this ability were the converging of author and reader, institution and audience, public realms with private spaces and other once-obvious dichotomies. Journalists, whose main job had been to help people navigate these societal divisions even as they perpetuated the status quo, must consider the convergence culture and its mass self communication characteristics and decide whether to adapt or not. This chapter will reveal the challenges and strategies that one transitioning newsroom encountered and developed as the organization attempted to forgo its comfortable print home of the newspaper and forge into the digital realm. At its core, the research examined whether the newsroom had adopted the principles of interactivity and multimodality, characteristics of this new culture.

At the end of this tale we will see that the convergence culture had only peripherally penetrated this newsroom at the time of America's information revolution, despite Jenkins' assertions of the culture's omnipresence. Ultimately this newsroom rejected some manifestations of the interactive capability of audiences and only haphazardly implemented digital media such as video and interactive graphics. Organizational norms that had been in place for the last century guided these journalists who had been trained to think of their role as a protector of society, a keeper of societal knowledge, and an agent of social power within this Midwestern capital city. And the journalists in this newsroom perceived that fully accepting digital-world practices and standards necessarily undermined the fundamentals of the traditional journalism they wanted to follow. Nonetheless journalists incorporated many key aspects of the emerging paradigm. And as a result, very early into the transition, they realized their long-held definitions of news and consideration of news values as well as their long-nurtured relationships with readers had to evolve. In that evolution, veteran journalists worried about the *CapTimes'* treasured identity in Madison and wondered if all of this "beaming up" would set up a situation where their authority as the main purveyors of credible, socially responsible news and information might be discounted without significant benefit to their constituency in the process.

An Ethnography

In February 2008, *The Capital Times* made the announcement that their afternoon broadsheet would be eliminated in favor of an online-centric product and two print weeklies. The moves were meant to help cut costs, stem the tumbling circulation figures and position the company to take advantage of the future. More than a third of the newsroom staff —some two dozen people —took buyouts or were laid off.[3] Most of the remaining reporters and editors had to reapply for their jobs, which were reconfigured for the new missions. A leadership team was formed, made up largely of those who had shown in the past to be technologically flexible or had experience in digital worlds. Several new staffers were hired: reporters, graphic artists and an online news editor. The average age of the staff went from about 50 to under 45.[4] By April of that year, editors were laying out the final print edition of the afternoon paper and the transition was complete. Or, rather, it had only just begun.

Right from the beginning of their transition, *The Capital Times* positioned itself as a "pioneer." In press reports and early interviews, the executive editor called the newsroom an R&D facility for both the company and the profession. They welcomed me, a former journalist-turned-professor, into their newsroom to document their trials and successes. During this time, the editors gave me full access to the newsroom as well as internal documents, emails, and memos to conduct this research. In all I spent about 240 hours or 6 full work weeks over the course of a year between 2008 and 2009, and was still in touch with newsroom staff as of 2010. In the true spirit of ethnography, I combined methods by relying on observation and conversation and then supplementing the data with both formal and informal interviews during this year (Jorgensen, 1989; Lindlof & Taylor, 2002). I interviewed just about all of the reporters and editors, as well as several marketing employees, the publisher, an editor emeritus, and former employees. I attended as many staff meetings as I could. I tried to remain an impartial observer. I agreed to keep all employees anonymous, share the data I collected and show them any manuscripts before publication. I ended up with hundreds of pages of documents, transcripts, and notes.

Using grounded theory, I approached the data through a series of coding protocols derived from Strauss and Corbin (1998) that included scrutinizing the material for dominant themes and concepts and then breaking those macro categories into sub-dimensions. A final analysis, called the selective coding process, helped contextualize the data according to a larger framework. For the purposes of this chapter, I sought information that related directly to the newsroom's framing of the transition and any reference to the normative methods for performing journalism, especially as they may have been evolving because of the medium change. I paid particular attention to references to interactivity and multimedia and the other distinctions of Jenkins' "convergence culture" or any nod to the emerging "mass self communication" that Castells had theorized.

The Transition

At the time of the newspaper's transition in mid-2008, the beleaguered industry buzzed with trends such as "journalism as process," which emphasized the "we the media" capabilities of the Web (Gillmor, 2004; Jarvis, 2009). *The Capital Times* recognized that to be "successful" (gaining readership and making money), its product must become both interactive and multimodal, or at least so the newsroom leaders hypothesized.

Interactivity. By 2003, Web 2.0 dynamics allowed a level of interactive communication online in that people could comment, share, and modify the information they were viewing (Downes & McMillan, 2000; Jensen, 1998). And as Web technology matured, the situation folded in Web 3.0 (also, the semantic Web and cloud networking) where content online could be organized, customized, mined, analyzed and otherwise manipulated and parsed by anyone in the sender-receiver transmission chain of communication. Newsrooms made only stumbling advances into this interactive world, with some embracing it and others preferring to watch it unfold at other publications. Eventually, most newsrooms had eventually adopted some kind of commenting system, instituted Q/As with readers, welcomed user-generated content and encouraged reporters' blogs, among other forms of interactivity (Emmett, 2009). However, most newsrooms maintained the status quo of the gatekeeper approach when it came to literal audience participation (Hermida & Thurman, 2008). Indeed, some scholars started referring to the "myth of interactivity" (Bergstrom, 2008; Domingo, 2008b) and criticizing Jenkins' "convergence culture" as too deterministic and as yet unviable and untenable.

Multimedia. By the time YouTube had attracted the attention of the populace in 2007, people were actively seeking out online videos, including news (Karbasfrooshan, 2010). The blogosphere and journalism trade publications heralded the marketability and journalistic importance of video and other multimedia for news organizations. Multimedia offered new narrative forms that could engage the reader with sensory experience so different from one-dimensional text. As one example, *Nieman Journalism Labs* published a piece about what journalists could learn from a post-feminist, sardonic Old Spice ad that went viral the summer of 2010 (Garber, 2010). Scholars investigated how video and other multimedia could help resolve some of the industry's credibility and audience issues because of its supposed transparency and its trendy popularity (Ibelema & Powell, 2001; Pfau et al., 2000).

This was the environment that the leaders of *The Capital Times* found themselves in as they realized they would need to do something to staunch their bleeding circulation figures. At the time of the transition, one editor acknowledged the challenge ahead:

> This newsroom was an extremely traditional one with an entrenched culture. But we have not been profitable for a long time. And we need to be relevant to our community. We need to figure out what the web site needs to offer beyond "news" exactly. It does seem a lot of what is out there is not about journalism but about getting newspapers to be information containers. How can *The Capital Times* play a role in all that? We needed a *CapTimes*-centric approach. We're blowing the place up and I don't know how you can do that without a little bit of dynamite.

Though the paper had long shoveled most of its stories online and had dabbled with digital technologies, little in the way of real convergence had resulted. Very few reporters even read the comments under their stories, participated in forums, initiated Q/As, recorded video, or conceived of interactive graphics. For the next two years, they focused on making the organizational changes necessary to promote interactivity and multimedia in order to become more "relevant" to their audiences. Unfortunately, the very notion of "relevancy" in this environment proved polysemic and problematic, as this chapter will show.

Definitions in Crisis

The challenges for the staff as it stopped the daily news product were many, complicated by a kind of joint-operating agreement with a sister paper/competitor the *Wisconsin State Journal*, a website called Madison.com that was run by a department owned by both organizations, and a decision to continue publishing a twice-weekly magazine (*The Capital Times* and an entertainment product called *77 Square*).[5] The tensions emerged right from the beginning. In this exchange, the editors discussed the wording of the headline the day the transition would be announced to the public:

> EDITOR 1: I like "Beam us up!" I just want to put the focus on the web.

> EDITOR 2: I think it is stupid because we don't want people to think we're going to some other planet. We're still going to be *The Capital Times*.

> EDITOR 3: I like it because it is "beaming" us up and we are actually going into cyberspace. "Here we go into something new!!!"

The short conversation foreshadowed the coming frustrations as well as the constant negotiation both within the newsroom and the community about the new mission of *The Capital Times*: Would they nurture the same identity from the printed version for the older Madison progressives or try to foster a new *Capital Times*, one more edgy and oriented to a younger audience? Immediately, reporters found themselves split between the two worlds of investigative magazine-style print-focused journalism that remained and the online world—both of which had to be learned for many of the daily writers. In the following section, I will detail how the transition forced re-conceptualizations of two key definitions for the staff: their identity as journalists and the product they were creating. The conclusion will explore what it meant to succeed in this new world according to the realities of the "convergence culture."

Redefining Journalistic Identity

In mid-2008, two editors were interviewing a candidate for a public affairs reporting position, peppering the young woman with questions ranging from her journalistic instinct to her skills in everything from photography to programming to socially mediated conversation. Said one editor after the interview: "Basically I want it all." "Wanting it all" reflected the convergence culture at its heart, including the converging of mediated tasks on a practical level but also the converging of roleplay for the reporters. Another editor expounded on the ideal *The Capital Times* reporter post-transition:

> I think about what needs to happen as morphing from being a reporter to being something that is an evolution of being a reporter, trying to give people the richest, best information in the most relevant way. Some of it is going to be reporting and some of it is going to be gatekeeping and some of it is going to be blogging and some of it is going to be video. I want them to be comfortable in all of those worlds. We need to be where audiences are. We need to invite them in.

Mass self communication capabilities combine the agency of the journalist with that of the audience (or author with reader and vice versa). This editor assumed that the journalistic "relevancy" included the presence of audiences with reporters, though actual productive agency rested with those recording video, managing content or blogging. The constant refrain of editors to "be relevant in this new world" engendered a feeling of uncertainty: 'relevance for what or who exactly?' they asked. And in the words of one reporter during a staff meeting, "Aren't we going to lose our identity if we do all these things?"

Relevance as a journalist in the convergence culture entailed significant task-management and job description reorganizational changes. Editors had to become marketers. Photographers had to learn video. Writers were asked to wield recorders and manage discussions. In one interview a writer complained of being an "art director" and a "computer programmer" as much as a reporter. "I have a lot more hats than I used to and I am willing to do that but I can't be everywhere all the time, and nobody seems to appreciate that part of this," said that reporter. Newsroom staff—most of whom described themselves as "noble writers"—reported feeling "off balance" and "anxious" about their obligations in the new world, and this anxiety was directly related to philosophical understandings about their professional identity. Consider the following comment after a writer experimented with video for the first time, converging his tasks:

> I took about half an hour of video and they are going to edit it down to about 5–10 minutes or so. It was so hard. I had to worry about audio and lighting, and it was in a bar and there was a lot of noise and I wasn't writing anything down at all!

Here the reporter grappled with his new identity as a multimodal reporter. His roles converged, for he was no longer what he once was (writer) but also a videographer, and he had to re-conceptualize the parameters of that identity. This led to significant labor concerns, including finding time to train in multimodal practices, being asked to multitask on the scene, changing expectations for production and rethinking the final work product completely. This kind of task turmoil (called "competence-enhancing discontinuities" by Tushman and Anderson) reflects a key and inevitable characteristic of technologically transitioning newsrooms, according to scholarship—turmoil that ultimately decreases over time and makes the company stronger (Cortada, 2006; Tushman & Anderson, 1986). "Convergence" in this instance referred not only to the medium change (print to video) but also to the job tasking and agency of the content producer—print reporter and visual journalist.

Much hesitancy surrounded these convergences. In the following comment, a mid-level editor in the newsroom cautioned the leadership team about entering the convergence culture too quickly, without consideration of the consequences to the professional identity:

> This is a real education for people who have not done this kind of journalism and do not know that you do not jump at the first press release or write the first blog that comes to mind. (We) are the filters for people. I think if we lose that vision and mission then we are not a journalism source any more. We have to do some educating about our worth and what we do and what we are compared to people who blog on their own time in their pajamas.

Here the editor reasserted the gatekeeping role for the organization, as the powerful agent over societal information in contrast to the blogger in his "pajamas." The term "pajamas" conjured dress for the non-work, private realm, and this particular phrasing relegated the "mass self communicator" to that of a nonprofessional, someone whose domain existed outside the public sphere. She discounted the product of mass self communication as journalism and reminded staffers to stay true to *The Capital Times'* original mission as a progressive filter of information for the public. Here was an instance in which we could see the tension between the organization's initial declaration to be the "pioneer" into the digital world where a "new and improved" *Capital Times* would live and the realities of the print institution's tradition, routine and culture that had sheltered the original *Capital Times* for nearly a century. Journalists resisted truly redefining themselves for the digital world even as they redefined their tasks.

Redefining News (versus Information)

As journalists struggled with their identity after the transition, they also grappled with whether a reconstituted news product was in order: What does news become in this converged world with its mass self communicative properties? Those in the newsroom found themselves conflicted over the implications of the mass self communication mode for news, as one short exchange revealed:

> REPORTER 1: Did you see all the nice comments on [Reporter 2]'s story?
>
> REPORTER 2: Yeah, I am just waiting for the nastiness to start.

Editors and reporters found the randomness and "irrelevance" (an oft-repeated term) of the Web culture disconcerting and contrary to proper information dissemination. Journalists in the observed newsroom countered the perception that they must offer interactive and multimodal content with preserving standards of news versus information.[6] In the following exchange, the editors debated the boundaries of audience participation within the site of the news organization—long considered sanctioned journalism content.

> EDITOR 1: There has been talk about distinguishing the forums from the news content. That is really important. Often I will go to Madison.com, and there will be these random links and irrelevant discussions. It's the Wild West.
>
> EDITOR 2: Although it is providing a platform for people to have a conversation. The idea is that that would be this vibrant community dialogue about whatever topics people chose to dialogue on.
>
> EDITOR 1: Yes, but it sends a mixed message. Much of it is offensive and then we are connecting to all of these other publications. It is confusing because it looks like we are endorsing whatever kind of blog they are starting....Why are we connecting to that?

Even as editors expressed a desire to foster a new kind of community dialogue, they wanted a careful distinction between what was journalism and what was not. Journalists separated their "news" content from external information found in other parts of cyberspace. They wanted separate domains for any user-generated content (reformatting all of their audience spaces to be completely separate pages from any news article). These actions were made in an attempt to preserve the credibility and authority of the reporters' authentic content as news. This supported what other scholars have found regarding journalistic concerns about user-generated content (Hermida and Thurman, 2008).

At the same time, the editors pushed the staff to "step outside their comfort zone"—breaking from their traditional conceptions of the "news article"—and yet some expressed confusion over what and how that product re-conceptualization would proceed. Said one editor: "Is it our impression that we are going to be able to be successful focusing the website on inverted-pyramid-type daily stories? Of course not, but trying to figure out what else to do is mayhem." Editors were asked for guidelines for the new world, but there were none to give as in newspaper days (One story a day! One Sunday story a month!). Reporters said they became over-whelmed in the constant pressure to "feed the online beast," as they called it, and have something fresh on the homepage constantly. One reporter admitted: "I do make one less phone call. I get it up fast, as they tell me. A lot of times I see us just putting up basically the press release. Is that news?" In this comment we can see how the concepts of "news" and information began to converge: what was once considered merely information—the stuff of press releases, police briefs and announcements—had become "news" that merited play on the website. Some reporters assumed that producing for online meant bulk content without interpreta-tion or contextualization so that audiences would have something to filter through.

Slowly, as the staff started getting used to their new beats, learning new routines and understanding the potential of the online platform, a shift occurred. More of the reporters bought their own digital cameras and asked for video help. Others executed blogs, helped the community-media editor tweet and mined social-media spaces for story ideas, sources and audience exchanges. An outside consultant was brought in to help guide the journalists toward interactive and multimodal visions of news. The consultant demonstrated the traditional journalistic value in such features and encouraged reporters to engage online in many different platforms, including with user-generated content. As the staff began dabbling in the interac-tive and multimedia tools, they reported finding value in digital media.

> Our survival depends on giving readers/users what they want and need and it does not mean that we cater to the lowest common denominator. It means that we determine what information is useful to people and how best we can go about that. We are still doing the essence of journalism. We are still the community's storytellers. But the audience is mobile and part of our job is to figure out where they are going and what they need next.

Said one reporter: "I love it because it's so immediate and we can help our readers... errr our viewers...know what is happening right now." Here you can see a recon-sideration of the audience from reader to "viewer," but no reference to the proactive audience member who might be self communicating online alongside the news article. It was not so much that the definition of "news" had changed for these traditional journalists as what forms it could take. To be "relevant" in the world of journalism, they needed to reiterate their own societal role as the information purveyor for society. They must remain "the community's storytellers."

Conclusion

This chapter began with the premise that Jenkins' convergence culture had created new information production dynamics that included "mass self communication." Mass self communication has been characterized by citizens' local ability to globally communicate in a mass manner, a new form of communication that fragments audiences because of integrated communicative platforms (Castells, 2009). This brief examination of the newsroom in transition demonstrated how a news staff attempted to account for the new capabilities of people to produce "news" across different media platforms. But in the process, this transition threatened to fragment journalistic identity and pervert notions of "news," at least in this newsroom. Was *The Capital Times* reporter a photographer or videographer or noble writer? Should editors design a product that could be significant to the mass self communicators of their audiences or to their traditional core constituency, the 50–60-something social progressives? Of course these roles and products did not have to be mutually exclusive, but carving out the time and resources to become competent and flexible in several different identities and for many different tasks to satisfy these conditions created tension among this staff.

The journalists in the observed newsroom found that as they adapted their kind of journalism for the web, they also had to reconsider what it meant to produce "successful" copy. A "successful" product entailed volume, freshness, connection and experience, according to the data from this ethnography. Furthermore, the newsroom underwent significant organizational restructuring. Reportorial labor, routines, and job tasks all had to accommodate the new platforms as producers re-conceptualized news beyond text and their fundamental "noble-writer" identity. Editors asked the reporters to invite the public into their production spaces and to expend effort to build relationships with *individual users* (as opposed to audiences). Newsroom staff was encouraged to accept former readers as authors, sometimes on the very pieces they considered to be sanctioned journalism. And finally, the converged culture for these journalists in this newsroom also meant revealing private newsgathering material in public worlds (either as blogs or in comments or as transcripts attached to their stories). One reporter blogged about his private, off-time activity (bowling); another (not a sports reporter) was tapped to blog a big baseball game from home on a Saturday.[7] An online editor created a couple of databases for the public.

Such features remained firmly in the control of the traditional mass communicator, however, with the journalists deciding what data to offer and how people could use it. Any "success" on the part of the newsroom staff had to relate directly back to the fundamental mission of *The Capital Times*, which was to provide socially responsible content as the "community storytellers." Convergence cultural "relevancy" in this newsroom came about only when organizational histories could be taken into account. The outside consultant, for example, connected all of the digital tools with traditional journalistic work. Said one editor, "There needs to be extra value to doing this video and not just shooting video because we are supposed to be shooting video. It needs to be worthwhile for the type of journalism we do at *The Capital Times*." The leaders needed to discover ways to adhere to the emerging digital dynamics while staying true to that essential core mission. They wanted to relegate any mass *self* communication of the convergence culture to the nonprofessional worlds and not allow the users into all of the journalistic spaces.

This evidence showed that Jenkins' convergence culture can be stymied by such organizational dynamics. For example, multimedia proved to be an almost insurmountable challenge for the newsroom as reporters had little time to learn such skills and still be successful in monitoring their beats. Even by mid-2010—the time of this writing—multimedia were virtually nonexistent on the site, save for the occasional outsourced video. No interactive graphics or multimedia packages existed for the most part. And, only a very few of the reporters engaged with their audiences within the commenting sections or their blogs. Thus, it is posited here that the true merging of worlds that Jenkins philosophized about cannot truly be achieved until the system itself has changed and perhaps the very definition of journalism (not to mention the poor fiscal situation facing most newsrooms).

It should be noted that digitization "amplifies the diffusion of the message beyond anyone's control....the network power exercised by digital networks assumes a new form: the removal of control over message distribution. This is in contrast with the traditional network power of the mass media, which reformats the message to be suitable for the audience in accordance with corporate strategy" (Castells, 2009, p. 418). Staying "relevant" while still maintaining control over the message may be impossible. This could not be tested in this ethnography, which centered on production practices of an institution and not on the audience. Audience studies and textual analyses of online content would shed more explanatory light on all of this. Still, Castells held gatekeeping power to be alive and well in this scenario, particularly among the government and media companies who grant or deny access to specific nodes within the network to other communicators. The editors and reporters at *The Capital Times* sought to maintain that filtering task in the 21st-century, fully transitioned newsroom. They did this through setting strict commenting policies, establishing social-media and other online protocols for reporters, and otherwise constantly reasserting their traditional goal as the authoritative "Wisconsin's Progressive Voice." The evidence from this ethnography undermined

Jenkins' contention that a true "convergence culture"—or at least its ubiquity—had taken hold in all of America's newsrooms.

ENDNOTES

1 This information comes from the executive editor of *The Capital Times*.

2 Though of course the organization did not use the specific term "convergence culture," editors made it clear in both external and internal communication that they sought to adhere to new digital dynamics.

3 The mandate to reapply for their own jobs created much angst among the veteran staffers. Though these non-union employees did not protest publically, their resentment certainly contributed to the initial difficulties editors had in transitioning the reporters to the new medium. As reporters got used to their new beats and the changes in general, the resentment abated. However, their uncertainty at the job expectations (as well as the stability of their employment—the company made four more rounds of layoffs over the next two years) continued.

4 Older workers tended to take the buyouts because of the lucrative severance package, tied to number of years with the company.

5 At the time of this writing in mid-2010, *77 Square* (including the entire staff) had been folded into the *Wisconsin State Journal*, so that it replaced the features' page once a week.

6 For more on reporters' motivations about user-generated content, please refer to Vujnovic's et al. (2010) interviews of online journalists and editors in ten countries.

7 It should be noted that both did so, willingly and happily.

CHAPTER FOUR

Redefining Public Radio
Marketplace in the Digital Age

Nikki Usher

Radio journalism has historically received very little scholarly attention. This chapter addresses the particularities of the development of online journalism in a newsroom that has audio as its core news material. The setting of the ethnography is *Marketplace*, the economy news service of American Public Media for National Public Radio in the USA. In the context of a production environment where journalists' main aim is an excellent radio product, the author analyzes the decisions of the newsroom shaping the production of podcasts or weblogs. The experiments with these new formats seek to conform with both the working routines of radio production and the philosophy of the newsroom, but are also an opportunity to explore new narratives and tones that end up being assumed in the radio production process as well.

EDITORS' NOTE

In the present crisis facing journalism, researchers' attention has primarily focused on the developments occurring to transform print newsrooms, and secondarily, on the changes taking place in television news. However, radio newsrooms are also undergoing considerable transformation as they attempt to ready themselves for the digital age. In the US, listeners are growing rather than receding and they are listening in new ways, most of which involve either the Web or mobile devices. Thus, radio newsrooms need to understand how to reach these audiences wherever they are because they won't always listen to audio news as regularly scheduled programming on a radio. As such, radio newsrooms have to work to build a bridge between their traditional radio product and the new ways listeners can get the information radio news provides—either through live streaming or via podcasts.[1] Moreover, radio news is different from television and print journalism in that it has no content that is naturally ready for the Web; raw content from radio is its audio product. This brings a unique set of challenges to radio newsrooms that print and television newsrooms do not face.

This chapter examines the particular challenges of a national radio newsroom in the United States as it attempts to develop these new bridges with listeners and potentially expand its reach. In 2009, I spent two days each week for five months conducting newsroom ethnography at *Marketplace*, the United States' largest business-oriented public radio program by audience. Senior managers were the only employees to explicitly state on multiple occasions to me in interviews that, "obviously the Internet and mobile are going to be a huge part of our future survival." This message was not repeated in the staff meetings I attended or in staff-wide emails listing *Marketplace* initiatives. Nonetheless, the actions I observed in the newsroom suggest that journalists understood the Web would be key to *Marketplace*'s continued relevance. However, the difficulty facing *Marketplace* staff was how to fit Web innovations into the daily cycle of producing its primary product: the daily, nationally distributed radio program. Here, I review how challenging it was for *Marketplace* to incorporate online journalism into the daily radio production clock despite the desire to do so. I also illustrate two examples of innovation that showcase *Marketplace*'s desire to move forward with new online initiatives: the Scratchpad blog and the podcast specifically created as new content, Dinner Party Download. Significantly, each innovation reveals a specific radio sensibility. Despite routines, journalists are self-reflexive about their work and the importance of the Web in the newsroom and make efforts to step out of the daily news clock to make room for these innovations.

Context and Setting for *Marketplace*

Though the US is not known for its public service media, it does have a vibrant public service radio sector. In 2010, the approximately 800 non-commercial radio stations received close to $420 million from the US government. These radio stations use this money and other money raised from listener and philanthropic donations to produce their own content, and more significantly, they act as "member stations" of national non-commercial public radio organizations from which they purchase programming of national interest. The largest of these US non-commercial public radio organizations is National Public Radio (NPR), and the second largest is American Public Media, the producer of *Marketplace*.

Marketplace is the US's largest business-oriented public radio show with a total of 9.1 million listeners, and it is distributed by 486 public radio stations (American Public Media, 2010). *Marketplace* actually consists of three shows, though the nearly 50 editorial employees work across each of the shows. First is *Marketplace* Morning Report, an 8 minute newscast that updates with new developments 7 times each morning Monday through Friday, during the morning "drive-time" when listeners are driving in their cars to work. Second is *Marketplace*, a 30 minute afternoon show that is produced after the stock market closes and runs at various points before, between, or after NPR's *All Things Considered*, which airs during the afternoon "drive-time." The third is *Marketplace Money*, which runs on the weekends and offers an hour-long broadcast that focuses on personal finance. This chapter is primarily concerned with *Marketplace*, the flagship afternoon show.

From its founding in 1988, *Marketplace* has been based in Los Angeles, intentionally a coast apart from the buzz of Wall Street and the group think of Washington. The goal, as explained to me by the Executive Producer, J.J. Yore, was to both take advantage of the rise of the Pacific Rim and also to do something that would be uniquely different. "One of the little aphorisms we adopted was a business show for the rest of us, which was designed to say we were a business program not trapped in the din of the East Coast mediacentric elite and [a show] focused on regular people."

Marketplace has a national and international reporting presence, with a vast network of staff and freelancers able to gather reporting material from virtually any place in the nation and the world. At the time of my research in January 2009–June 2009, *Marketplace* had major bureaus in New York and Washington, D.C. in addition to Los Angeles, where my research was conducted. There were also one-person bureaus in the Pacific Northwest, Florida, North Carolina, Shanghai, and London, though since my departure there has been some shifting of this geographical distribution. There is a fairly even gender distribution in the newsroom and a wide range of ages represented, but the most prominent positions tend to be held by middle-aged people. The staff is mostly white, though there are people of color in top editing positions.

Methodology and Theoretical Framework

I spent two days a week, five to eight hours a day, for five months at *Marketplace*. My research involved observing meetings and shadowing reporters, editors, and producers so I could observe their work practices. Fifteen follow-up interviews were conducted; these also included interviews with those in the newsroom with whom I was unable to shadow. I also worked at *Marketplace* as an "intern" on one day a week, where I was charged with sending out the afternoon show's daily rundown—

the "DACS," or the news budget—to all of the member stations across the public
radio spectrum. I was given an email address and access to staff email lists.[2]

This paper takes its theoretical cue from some of the older ethnographies of
newsroom production practices, most notably, Gans (1979), Tuchman (1978),
Fishman (1980) and Epstein (1973), who set the tone for newsroom ethnographies
by spending years watching work practices unfold. Though their findings depict
how journalists shape and create news, their findings suggest a concept of newswork
that is based on routines: news needs to be produced for a timeline in order to get
the product out, reporters and editors rely on sources for adequate content, and the
influence of external forces such as the audience and profitability may not shape
content directly but influence decisions about news prominence and its accessibility.
Significantly for broadcast news, Schlesinger's (1978) idea of "stop-watch" journalism is particularly relevant: he argues that journalists are slaves to the production
clock and make decisions about what to cover and how to cover it based on time
and in addition, commercial relationships. There are few ethnographies about radio
newsrooms (Eliasoph, 1988), but Brannon's (2008) conclusions suggesting significant
similarities between broadcasters' online operations and other online newsrooms
provide some insight into the inner workings of NPR as it struggled to adapt to
the needs and practices of online news.

But my study of *Markeplace* reveals a newsroom that is much more creative
than has been accounted for in past literature. I wish to extend the thinking of
Cottle (2000) and Domingo (2008c), who critique older ethnographies because
they suggest we should pay more attention to the way human actors contribute to
and think about technological innovations in news production. Further, I argue
that within journalism routines, there is considerable opportunity for self-reflection
about newswork that contributes to an evolving news environment. In this way, I
argue that while routines exist, the discussions and debates that begin each day
represent new challenges—and the confrontation with technology signals a chance
for creative disruption, even if this does create new routines.

Online Journalism in a Radio Newsroom

Too often it is easy to think of online journalism as perhaps in its dominant form
as a text, photo, or video product produced by what were once primarily print
newsrooms now in various stages of convergence. When broadcast newsrooms have
been studied, such as they were in the previous volume of this book by Brannon
(2008) who reviewed ABC News, NPR and *USA Today*, the emphasis has been on
similarities—what these newsrooms all shared in adapting to online journalism.

However, online journalism for radio newsrooms is different, and creating it
presents distinct challenges. Deuze (2004) agrees, noting that there is considerable

argument over what constitutes multimedia journalism, and by extension, online journalism. First, audio newsrooms produce no naturally occurring text that can easily be placed online. Newspapers can simply put text stories up online after they have written them. But audio scripts are not written as news stories. *Marketplace* has no reporting staff to turn these audio stories from sound into easily navigable narrative text stories, and there is no plan to add staff to do this.[3] The solution has been to simply put up the transcripts of these pieces, but these can be unwieldy and may not make sense to a reader. Unlike broadcast news and print news media, there is also no naturally occurring output of visual content; at public radio networks and stations, including until recently at NPR, there is no previously designated staff to take pictures or produce video content. Blogging is more complicated for an audio newsroom; audio staffers are not used to writing in a way suitable for the eye rather than the ear, even if blogging may be a more conversational Web genre.

Podcasts are especially important when considering online audio. It would be foolish to consider podcasts as simply another means of distribution of audio production. Radio networks and shows are creating efforts to produce unique audio content specifically for the Web as well as for other listening formats, such as mobile. As such, one of the aspects that will be examined here is the way that podcasts have been re-imagined as a form of online journalism for *Marketplace*.

Marketplace is an organization grappling with how to adjust its daily radio broadcast routine for what is perceived as a new and necessary form of engagement with the audience. One editor said to me when crafting a project that required online participation from listeners, "This is how we are going to reach our audiences now." But the broadcast clock makes online journalism difficult for *Marketplace*. An analysis of two typical moments in *Marketplace*'s daily routine—the morning meeting and a reporter's daily work process—reveals how on the broadcast clock, online journalism becomes quite challenging.

The Morning Meeting

A typical day at *Marketplace* begins with an 8 A.M. story pitch meeting. At this meeting senior editors and producers begin by listing story ideas. Principally, this meeting is intended for the "newscast" portion of the afternoon show—the six minutes (sometimes longer) at the top of the hour. Occasionally, these meetings will generate a longer story for later in the newscast which may be given anywhere from 2–4 minutes, but this is principally devoted to a discussion of hard news.[4] This morning meeting is principally intended to set in motion the day's radio production process.

When I arrived in January 2009, *Marketplace* staffers were trying to deal with the fall-out of the financial crisis on banks and ensuing regulatory efforts, and were attempting to explain what the recession would mean to average Americans. An oft-repeated refrain at news meetings by one of the hosts was, "Why does my 69-year-old mother care? And will she understand?" These questions suggest that in the routine of the meeting, there was a self-reflexive awareness about the place of the audience in their work.[5]

The senior staff attends the meeting, along with the intern and the Web producer. Stories germinate from a variety of sources—trends observed through the daily life of *Marketplace* hosts, producers and editors, major newspaper outlets, and cable and broadcast TV. The goal was to find stories that could be told in a "Marketplacey" way: a way that spoke to the audience that was not interested in traditional business news.

On January 15, 2009, just days before President Obama's inauguration, a number of stories were in the mix. The news meeting began in a way that was indicative of many news meetings: The senior producer announced the nine-minute news hole (the rest of the show would be filled with interviews, commentaries, and feature segments). The morning host began rattling off the day's news.

Dozens of story ideas were considered from Vitamin Water having an initial public offering to Microsoft layoffs to the Sundance movie festival, all with editors asking what the *Marketplace* angle would be on these stories. "What do these Microsoft layoffs mean to the rest of us?" one editor said. "Isn't it just more bad news about the economy?" A spirited discussion about Bank of America ensued because editors and producers agreed this was the big news of the day—the news listeners would really care about because of the significant government intervention. The senior producer asked, "What are we getting for our money? This is our investment?" and then the Washington editor noted, "What's our stake in all of this?" They then decided to do two stories related to the issue—one on Bank of America, and the other on how Bank of America was not the only bank facing financial trouble.

This editorial conversation shows that editors and producers were actively trying to ask a different set of questions that would inform listeners outside of the Wall Street–NYC corridor about "our money." While the news meeting itself is a routine, the conversation showcases spontaneity and creativity. Earlier ethnographies have failed to capture the dynamics of these conversations. But these conversations reveal that within routines there is flexibility and room for new ways of thinking about the meaning of stories. In defining what counted as a *Marketplace* story, these staffers were engaged in a self-reflexive process about the organization's values and its identity.

Immediately following this meeting, the intern and Web producer were instructed to have online, within 15 minutes, three blurbs about what would be up on the show. These would be short briefs about what would later develop into a full online presence by approximately 2 P.M. LA time or later when full scripts were in and edited for the website. The only images for these blurbs would come from stock photos from the Getty agency. In this case of January 15, the online summary early on looked like this:

> With Bank of America seeking more TARP money to ease acquisitions of smaller competitors, can we expect banks such as J.P. Morgan Chase and Wells Fargo to request more bailout funds as well?

This brief did not represent *Marketplace*'s witty take on a difficult business story. In the time crunch of production, the Web producer and intern could only work from cues from the meeting and from quick conversations with editors and producers about each story. Further, and perhaps more significantly, the story had not been crafted yet; the *Marketplace* angle had yet to be refined.

Nonetheless, those in the newsroom insisted on the importance of getting these briefs out as quickly as possible. As an editor explained to me on my first day, "If we get these [Web blurbs] up quickly we can build anticipation for the show later." Thus, two elements limit the online content: the first is the uncertainty, the Web staff simply doesn't know what the story will be when the blurbs are published; the second is the pace, as Web producers have just 15 minutes to launch Web content before the rush begins to meet the show's 1 P.M. deadline.

A Reporter's Daily Work Process

The pace of news radio reporting is gruelling and makes it difficult for individual reporters to contribute anything online. Stories were assigned at 9 A.M. for a show that began recording at 2 P.M.; the stories had to be mixed and produced to fit into their time segment by 1 P.M. From a technical perspective, most of the day is filled with reporters, editors and producers trying to make the sound as high quality as possible. Because the reporters are already under two types of time pressure (fitting the story into an assigned time segment and making a 1 P.M. deadline), adding additional online elements into the routine was difficult.

Marketplace content is handled in a horizontal manner involving considerable teamwork. Assignments are either handed out by the editors or are suggested, particularly in the case of feature stories, by reporters themselves. Editors work with reporters to build the scripts though reporters write the first version. Reporters record soundbites using specially adapted phones or digital recorders if they interview their sources on the spot. They then use a digital editing and recording system to get the pieces that they want. Then, they go into a studio to record their own voices and mix these soundbites into the quotes or ambient sound they have gathered from their reporting. The best sound comes from phone connections that are directly connected to a digital audio line, such as the kind of line that can connect two studios; cell phones are avoided, and in the case of feature stories, phone interviews are expressly forbidden—sound must be gathered on the spot.

In March 2009, I shadowed a reporter who was covering the downgrading of General Electric's bond rating. The idea of the story was to try to explain how GE was under attack and what that might mean for the future of the company; the reporter had 2 minutes on air to do the job. The reporter began by trying to isolate sound for GE from a CNBC appearance,[6] working with the production assistant to get permission to do this. The reporter then began leaving calls with a number of analysts, noting, "I'm calling from *Marketplace* public radio, this is '…' and I'm working on a story today in Los Angeles. I spoke with you on a story about GE and we are working on a tight deadline." After calling about five or six sources, he managed to get in one in addition to the GE cut from CNBC before writing his lead before 12 P.M. During this time he digitally edited sound from a 15-minute interview to get 15- to 20-second cuts from the interview, noting "10–12 seconds is ideal."

He also had to, in his words, "cut tape" from the CNBC interview using electronic editing software to chop out sound clips to use in his story. The challenge was to compress a vast amount of material into a short amount of time, and the reporter had to record the script by about 1 P.M. He noted he couldn't get everything into the story, "I just couldn't get into the second half about what it meant to investors. I ran out of time." This reflexive moment reveals that the reporter was actively thinking about how he was trying to tell the story to his audience. While radio news production is limited by the clock, journalists are aware of the strengths and weaknesses of their work because of these routines, suggesting that news production is an evolving process where journalists see continual room for refinement in their work. But once again this reporter was limited by the clock from contributing to the website. There was no time during his day when he wasn't calling someone on the phone, editing audio, writing a script or recording for the radio before 1 P.M. for him to think about adding additional online elements.

Online Journalism Opportunities

The Scratchpad Blog

Off the clock, however, *Marketplace* can find ways to innovate online. *Marketplace* made a deliberate venture to expand its online offerings with the introduction of Scratchpad, a new *Marketplace* blog, in mid-February 2009. The goal was to make this blog "*Marketplace* in blog form"—to replicate online the core values of the radio program. In fact, the blogger was given the title of "Online Host." To underscore this point, the blogger was one of the former hosts of the morning show, Scott Jagow. In this symbolic move, *Marketplace* showed that

Scratchpad would be an extension of its audio offerings but in online form. The blog remained closely connected to the rhythms of the audio newsroom; Jagow attended the regular morning meetings before deciding on his orientation for the day and continued to offer story ideas. However, as "online host," Jagow had none of the restrictions of the broadcast clock. Instead, his goal, as stated in the introduction to the blog, was to be additive and experimental and to offer more than what was going on the radio.

Shortly after beginning the blog, on March 18, 2009, Jagow "channeled" his anger at the American Insurance Group by making his own movie on the Web about the company, called "AIG Performance Review."[7]

This video hit home with those at the morning meeting as they began to see what the blog could do to extend the presence of *Marketplace* online. Staffers told Jagow, "That video was really brilliant" and "I forwarded it around" and "That was just hilarious, how did you do that?" Jagow had accomplished two feats with this video; he demonstrated he could be a "host" outside radio with his own voice, and he illustrated to people in the newsroom what the blog could do. Though *Marketplace* offers lighter stories in its broadcast, its coverage of hard news remains quite serious. The blog gave journalists a space to bring some of this humor and and context to subjects they were covering as serious hard news day in and day out, such as the corruption, frustration and difficulty with the government's bailout and rescue of the American Insurance Group. With the blog, people saw change, but they also saw a chance to redefine the identity of *Marketplace* in an online environment. Before, *Marketplace* online was just a collection of story summaries that had been posted by the intern and associated audio. Now, the site had a dynamic host who was actively involved in giving *Marketplace* online a personality.

The blog began to reflect more involvement from people in the newsroom. For example, by the next week, on March 24, 2009, the host of *Marketplace*, Kai Ryssdal, sent Jagow a YouTube video called "Recession Rock" that became the subject of a Scratchpad post. Ryssdal was making sense of the new potential of the Scratchpad blog and finding a way to incorporate it into his daily work.

Scratchpad also became a vehicle of newsroom transparency for listeners of *Marketplace*. Inside the newsroom was a "cliché wall" of phrases for producers, editors and reporters to avoid using when talking about the recession or the economic fallout. Jagow wrote about this cliché wall, which included terms the newsroom needed to avoid when talking about the economic fallout such as "Main Street, not Wall Street" and "You are not alone," and "X,"[8] giving the audience insight into what people in the newsroom were struggling with. This post was a moment of sensemaking for Jagow—he realized his position as blogger and saw that he was an intermediary between the audience and newsroom rather than simply someone putting content directly out for the audience without their feedback. He continued to engage in this role by actively responding to user comments. Jagow began receiving "callouts" on-air to direct traffic to his blog, another indication that staff was aware of the way an online presence could add to *Marketplace*'s audio broadcast.

Despite the promising progress of Scratchpad, the challenges of producing online journalism continue for *Marketplace*. Jagow's work for the blog was funded by a year-long grant from the Corporation for Public Broadcasting. *Marketplace* could no longer support the blog as of March 2010, and the experiment ended. It is now experimenting with a blog that includes voices from many staffers.

Experiments in Podcasting

One of the key venues for *Marketplace*'s expansion in the digital realm is podcasting. Some of the younger staff members were given time off the broadcast clock to create a new podcast called "Dinner Party Download: The show that helps you win the dinner party." The goal of this download, to be produced by 2 P.M. on Friday afternoon, was to offer to listeners the kind of things they might want to discuss at a dinner party, where so often people might say, "I heard this on the radio," or referring to public radio in the US, "I heard this on NPR." The download would be short and quick—a highlight in an amusing and more playful tone than some of the hard news content of *Marketplace* (though *Marketplace* does pride itself on quirky stories and accessibility). As one of the originators of the show told me while prepping for it one Friday, headphones in hand while editing the segment, "It's supposed to be fun and funny and more geared to a younger audience. To liven things up a bit."

The Dinner Party Download began in July 2008 as strictly online-only content with its own Web presence and was about 13 minutes long, some of it only vaguely connected to business, such as a tale about John Dillenger, the old-time bank robber, Easter eggs and Obama's three-pointers. This was an example of the unique content *Marketplace* could offer in audio form online.

But as the Dinner Party Download became increasingly successful, the senior producer began to notice that it could also become an integral part of *Marketplace*'s on-air show. "It's light, it's funny," she told me. "It's a good end to the week." This comment from the senior producer showed that *Marketplace* the radio program could change itself in response to what it offered online, and each could inform the other. The Dinner Party Download would remain its own 13-minute segment for podcast, but it was also available to listeners of the terrestrial *Marketplace* broadcast. Quickly, each Friday, the DPD as it was abbreviated, became part of the routine. On its second debut, the senior producer announced it as just another part of the show at the morning meeting—underscoring how routine was both flexible and amenable to change.

Dinner Party Download would now be aired every two weeks as a close to the show. The podcast, however, would appear weekly. These younger staffers were given permission to take time away from other stories and projects in order to produce this podcast, a sign that the newsroom recognized its value. This was a key moment in developing new routines to negotiate how online journalism would intersect with the traditional radio product. Management and the journalists working on the podcast noted that the online component could offer new elements to the radio show, but it could also exist on its own as a valuable complement.

For the senior producer and for those at *Marketplace*, Dinner Party Download was not talked about as a marker of change, but it was seen as something new and different—as "small talk" that could help add color to the show and could create a more youthful-sounding end of a long week of business news. The ultimate decision to make the change for the program came from the producer, but the idea to include a new media sensibility came from younger staffers who saw that they could reach new audiences online through their podcasts. Altogether, the newsroom realized the potential that new media with an audio sensibility could add to *Marketplace*'s value for its listeners.

Conclusion

The broadcast clock of *Marketplace* made it difficult for journalists to bring online journalism into their daily routines because so much of *Marketplace*'s efforts went into creating the daily show. However, these routines were not inflexible and could be adapted to include online journalism. The Scratchpad blog and the Dinner Party Download represented two instances in which *Marketplace* staffers became aware of the new opportunities that the Web could bring to the existing brand of *Marketplace* journalism. Notably, within the routines of daily news production, *Marketplace* staffers are actively engaged in trying to understand what their journalism means for their listeners. At *Marketplace*, routine and self-reflexivity meet; daily practices are considered and evaluated. The awareness that innovation online needs to occur outside daily pressure is perhaps the most significant indicator that these journalists are actively considering what their work traditions and work practices mean as they adapt to the online world. Radio newsrooms have unique ways that they can contribute to online journalism—through podcasting and through the distinctive sensibilities, such as a having an "online host" as a blogger. The intimacy that many listeners feel in the relationship they have to a radio host may signal yet additional ways of developing the audience/listener experience online that have not yet been explored by *Marketplace*.

At *Marketplace*, a new place for humor and and increased opportunity for listener-engagement and newsroom transparency were the result of online innovation. However, *Marketplace* was limited in its ability to change to advance its online newsroom. The opportunities for incorporating online journalism, particularly writing for the Web, remain lost in the broadcast clock and routines of *Marketplace*'s daily production routines. With Scratchpad, a single personality was crafting a new relationship with the audience rather than *Marketplace* staffers themselves taking daily efforts to using online journalism as part of their daily work. Time for innovation, such as the Dinner Party Download, required the creation of new routines and time away from daily newsroom pressure. Thus, *Marketplace* is still struggling to redefine what it will be in the online world and how its journalists will contribute. However, the daily newsroom practices do show a newsroom capable of evolution—if one step forward, two steps back—and journalists who think creatively and self-reflexively about their work.

ENDNOTES

1 Live streaming refers to audio or video that is broadcast in real-time over the Web from a content provider; podcasts are audio downloads of files that can be played on computers or on mp3 players.

2 Due to my agreement with those at *Marketplace*, in exchange for access, I do not name any participants unless they can easily be found on first reference to their work or their position by a simple Internet search.

3 NPR, however, does have a more robust digital staff that crafts separate digital stories, some to accompany radio pieces and some that are completely original.

4 Hard news in the business context refers to breaking stories about economic conditions such as changes in economic policy or major daily news developments in domestic and international finance, technology, manufacturing and labor sectors.

5 When asked about the audience, the majority of *Marketplace* staffers described their understanding of the demographic profile of the audience provided to them by APM: the majority of listeners between 25–54, the next highest group over 55, mostly upper middle class, white and college educated with a fairly even gender split. However, they would often counter this profile with experiences they had talking to truck drivers and sales clerks about *Marketplace* having meaningful stories for them.

6 Executives from Dow Jones Industrials 30 companies will rarely appear on a cable business news channel such as CNBC proactively, and such a move was at the time considered a sign that the company was under duress and needed to act aggressively to reassure investors, the media, and the public at large.

7 http://www.youtube.com/watch?v=_avTNR2THTM

8 http://www.publicradio.org/columns/marketplace/scratchpad/2009/02/only_time_will_tell.html

CHAPTER FIVE

The Internet in the Print Newsroom
Trends, Practices and Emerging Cultures in Zimbabwe

Hayes Mawindi Mabweazara

The evolution of news work with the adoption of the Internet goes beyond online and converged newsrooms. This chapter focuses on the work practices of newspaper journalists in Zimbabwe and explores the role that the Internet plays in their information gathering and relationship with sources. From search engines to social networks and email, journalists naturalize the convenience of online tools with practices that try to minimize the ethical risks they perceive. The singular context of an African media system dominated by state-controlled newspapers adds an extra value to the Internet as a window to alternative points of view on the events journalists report. The author argues for the need to immerse in journalists' work context in order to understand the reasons and implications of their use of digital technologies.

EDITORS' NOTE

The advent of new technologies (ICTs) in Africa in the 1990s "sparked celebratory, almost utopian bliss" among their proponents (Banda et al., 2009, p. 1). It was accompanied by the hype about the continent's possibility of "leapfrogging" some stages of development (ibid.). In the context of journalism practice, new technologies were seen as having the potential to increase journalists' work efficiency and thus overcome the barriers associated with the 'traditional' means of journalism practice. Research on African online journalism has equally been informed by various academic and practitioner accounts rooted in "*technological myopia* and *technological determinism*" (Cottle & Ashton, 1999, p. 22, emphasis original) which overlook the differential political, economic, cultural and social circumstances in which the technologies are harnessed and assimilated.

This chapter is therefore an attempt to contribute an African, particularly Zimbabwean, perspective on mainstream newspaper journalists' use of new technologies in newsmaking practices. The chapter departs from deterministic approaches and reinvigorates traditional sociological approaches to both journalism and technology to closely examine how Zimbabwean mainstream print journalists in the state-controlled and private press deploy the Internet and email in their everyday professional practices. The study is premised on the view that to understand the impact of new technologies on journalism practice in Africa, "we must put journalists into a critical and analytical context and begin to question the immediate and wider social context" (Mabweazara, 2010, p. 12) in which they deploy the technologies.

This approach finds root in the collective strengths of two broad theoretical concerns: the *sociology of journalism* and *social constructivist* approaches to technology (see Schudson, 2000; Bijker, 1995). Although these theoretical bodies were conceptualised before the 'new media age'—in the 1970s and 80s—together they provide a basis for conceptualising the interplay between journalists, their immediate context of everyday practice and the wider social factors that coalesce to structure and constrain the deployment of new technologies.[1] Against this backdrop, the chapter explores the patterns and trends in the use of the Internet and email by print journalists as well as their impact on practices and professionalism. Although practically these technologies are inextricably connected, in this study they were treated separately for analytical reasons.

Research Context and Methodology

To gather empirical data for the study, I deployed a triangulated ethnographic approach involving the use of multiple cases, multiple informants and multiple data collection procedures as shown in Figure 1. The study comprised six newsrooms drawn from the dominant state-controlled Zimbabwe Newspapers Group (two dailies, *The Herald* and the *Chronicle,* and two weeklies, the *Sunday Mail* and the *Sunday News)* and the small but vibrant private weeklies owned by Alpha Media Holdings (*The Zimbabwe Independent* and *The Standard*). The newspapers are located in Zimbabwe's two major cities: Harare, the capital, in the North and Bulawayo, the second largest city, in the South.

Following the increased use of the new technologies as journalistic tools throughout the developed world, the selected newsrooms took the lead in harness-

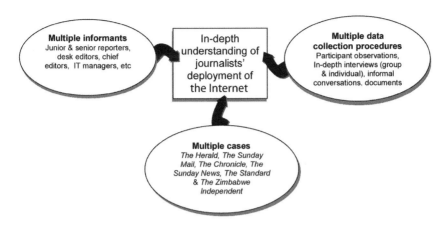

FIGURE 1: Multi-method triangulated approach

ing the perceived power of new technologies in their newsmaking operations. All the newspapers assumed an online presence in the late 1990s. In 2007 *The Zimbabwe Independent* and *The Standard* refurbished and modernised their websites with interactive features that enable readers to contribute content to the newspapers. The weeklies sought to use their websites to get leads and insights from their readers in the wake of the intensifying political impasse at the time (Moyo, 2009). The newsrooms also intensified their use of email to reinforce their links with sources (and readers) and for content exchanges with civic organisations such as the Crisis in Zimbabwe Coalition and Kubatana.net. In general, these newspapers' adoption of new technologies in the late 1990s constituted the formative stages of online journalistic activity in Zimbabwe. However, online news production in its more defined form (characterised by standalone news websites), only emerged around 2000 as part of initiatives by exiled Zimbabwean journalists.[2]

Overall, 96 journalists were involved in this study either through observations (including informal conversations) and interviews at diverse locations in Harare and Bulawayo. The research's thrust was not the generalisability of cases to populations, but rather the generalisability of findings to theoretical propositions (Mason, 2006). In total, the observations and interviews were conducted over a period of 157 days between June and December 2008.[3]

New Technologies in Everyday Practices and Routines

In Zimbabwe, the Internet and email have become essential tools in the print journalists' day-to-day newsmaking routines. The technologies facilitate the communicative processes at the heart of the newsmaking practices and play a significant role in helping journalists to move beyond the confines of traditional newsgathering methods. To highlight the centrality of new technologies in the newsrooms studied, we can take the example of a reporter from *The Herald* working on a story assigned to him by the news editor in June 2008.

One morning after the 'diary meeting'[4] I decided to 'shadow' a senior health reporter assigned to work on a story about the alleged decrease in HIV/AIDS prevalence in Zimbabwe. As a starting point, the reporter decided to use the Internet for background information and a general feel of the circumstances behind the alleged decline in HIV/AIDS prevalence. As we were chatting, I noticed that he constantly switched from one website to another and back to the page in which he was typing his story. In fact, he constantly copied material from some of the sites and pasted it into his story.

After a while, clearly satisfied with his progress for that moment, he reached for his mobile phone, scrolled through, and scribbled a number in his notebook and walked towards the news editor's desk to make a phone call.[5] After a few minutes, he returned to his desk complaining that the number he had been trying was incessantly on voicemail (it turned out that he had been calling the Minister of Heath and Child Welfare for comment).

As though unfazed by the unsuccessful call, he skimmed through a box on his desk and flushed out a business card belonging to Tim Hallet, a Research Associate at Imperial College in London (who, as I was to learn, had at some point partnered with the Zimbabwean government on HIV/AIDS research and the reporter under observation had met him on one of his visits to Zimbabwe). He instantly set out to write Tim an email, but before he could type anything, his mobile phone rang and he began a lengthy conversation. Immediately after the call, he revisited his story and remembered to send Tim Hallet an email. By this time it was almost 12.00 noon, lunch hour in this newsroom, so we set off for the staff canteen....

Immediately after lunch, we returned to the reporter's desk and, intuitively, he opened his email and noticed with excitement that Tim Hallet had responded to his email. Unfortunately, it turned out to be an automatic 'out-of-office response.' Without wasting time, as the deadline was drawing closer, he went back to his story, added a few lines and forwarded it to the news editor—exclaiming "that's done!"

The practices described here establish a point already made about the centrality of new technologies in Zimbabwe's mainstream press. Although it does not sum up all aspects of the technologies' functions in the newsrooms, it is telling in terms of how the technologies are enmeshed in daily newsmaking routines. From the narrative it is clear that the Internet is a key research tool that has also enabled reporters to directly connect with their sources through email. However, it is also clear that new technologies have not supplanted the traditional forms of communication such as the landline phone.

Also evident in the narrative are the ethical and professional challenges that have emerged with the use of the Internet in the newsroom, precisely the practice of 'copying and pasting' material from the Internet (sometimes without attribution, verification or attempts to talk face to face with sources). Against this background, the sections that follow explore the patterns and trends of the deployment and appropriation of the Internet and email in the journalists' day-to-day newsmaking routines and their impact on practices and professionalism.

The Appropriation of the Internet in Newsroom Practices

One of the key findings of the study is the centrality of the Internet as a source of news and story ideas for newspaper journalists. This function was closely tied to the general challenges Zimbabwean mainstream journalists face in sourcing news due to the polarised political context in which they operate. According to Mukundu the contextual challenges faced by the Zimbabwean media entail that newspapers undergo "unique newsroom processes and routines, both in order to survive, and in order to produce quality news products" (2006, p. 28).

The centrality of the Internet as a news source, however, follows the rise in foreign-based Zimbabwean news websites run by exiled Zimbabwean journalists such as Newzimbabwe.com, Zimonline.co.za, Zimbabwesituation.com. These online newspapers enjoy "'unfettered liberty' to publish what they want" (Chari, 2009, p. 23) hence always set "the agenda for the mainstream media in Zimbabwe by coming first with the stories" (Moyo, 2007, p. 91). As one senior political reporter at *The Zimbabwe Independent* explained:

> The pervasiveness of Zimbabwean news on the Internet spurs us to keep track of online news especially news websites run by our former colleagues who have sought refuge in neighbouring countries and abroad. We want to keep abreast of general news and current affairs, but also avoid giving our readers (the majority of whom read news on these websites) stale news.

The Internet has thus become a central tool for the mainstream journalists to access and explore news especially on topics that are not covered by the local press because of the constrained nature of the context in which they operate. Writing from an American context Pavlik (2001, p. 33) observes that "[t]he Internet…provides access to news and information otherwise often censored by governments and others in control of traditional media."

Beyond news sourcing, the Internet has also helped to overcome some of the barriers imposed by traditional research approaches. As shown in the narrative of my newsroom observation above, online research is widely used for background information as well as to seek clarifications on specialist topics. Emphasising the importance of the Internet as a research tool for journalists, Machill and Beiler (2009, p. 201) highlight that with the Internet "even editorial offices with poor research capacities now have at their disposal a relatively good means of conducting research."

As with the health reporter in the narrative given earlier, a number of journalists highlighted that they use the Internet as a research tool when they want to tackle specific assignments as well as learn from other journalists. As one reporter at the *Sunday Mail* put it: "English is not our first language, so when you read the UK *Guardian* or the *Telegraph* online, for example, you try and learn or even copy their writing styles...." In support of these findings, Pavlik (2001, p. 77) notes that "[t]he Internet is a useful resource for professional development and continuing education for journalists." Of significance, however, is the role of the Internet in facilitating Zimbabwean journalists' continued reliance on Western journalistic forms as a measure of their own practices. This highlights the fact that newsmaking practices in Africa continue to be influenced, at least at a general level, by Western patterns of newsmaking (see Nyamnjoh, 2005).

The deployment of the Internet as a research tool was, however, not without practical challenges and ethical implications. From observations and interviews it emerged that a number of journalists had difficulties or limited skills in exploiting various online search options including using topic-specific search engines; evaluating the quality of online content; and properly attributing material taken from the Internet. This scenario pointed to "a moderate level of [Internet research] competence on the part of journalists" (Machill & Beiler, 2009, p. 201).

Beat Influences on the Use of the Internet

Although there was a general trend in the use of the Internet across different news desks, the beat system also shaped and influenced how journalists deployed the Internet. This found explanation in the fact that each beat requires "very different coverage routines, different work rounds individually tailored to the particular activities within the beat's territory" (Fishman, 1980, p. 36). A more striking pattern of the use of the Internet was evident among entertainment reporters. A regular observation of Web browsers in the newsrooms indicated that entertainment reporters invested more time on social networking sites than their colleagues in other beats did. Facebook was particularly seen as a valuable platform for connecting with potential sources scattered across the globe. One entertainment reporter at the *Chronicle* spoke passionately about the importance of Facebook:

REPORTER: I do some of my stories through Facebook. As you know artists and entertainers are scattered across the globe, so I have befriended most of them on Facebook and sometimes when I do stories I interview them using the platform...

RESEARCHER: Any examples of stories that you have done through Facebook?

REPORTER: Yes, I have just done a story on Arthur Mafokate, a South African musician through Facebook...

RESEARCHER: But have you ever met him in person?

REPORTER: No, but the good thing is that I have pictures on my profile, so he has an idea of what I look like...[Digresses as he directs me to his computer monitor], look this is the story I am talking about...

RESEARCHER: So all these direct quotes in this story were sent to you through Facebook?

REPORTER: Yes, they are from our chat on Facebook word for word. Arthur actually sent me pictures of his child through the same platform. What made it easy for me to link up with him is that I am a friend to some of his friends on Facebook, and I have interviewed some of them for stories...so that guarantees my legitimacy to him. Sometimes you can actually pick stories from Facebook....For example, if an artist writes something like: "I had a boring show" on his or her profile, that is a scoop on its own! You quickly send a message to ask how big the show was and what went wrong and so on...

The example above shows that social networking sites are not without journalistic value for entertainment reporters. It also illustrates that the functions of social networking sites are not 'fixed' as technological determinists would suggest, rather they are a site of interpretive work that is entrenched in the intricacies of social interactions. This reinforces Bijker's (1995, p. 6) view that: "[O]ne should never take the meaning of a technical artefact or technological system as residing in the technology itself instead one must study how technologies are shaped and acquire their meaning in the heterogeneity of social interactions."

The utility of social networking sites as journalistic tools was, however, not without challenges. In reflecting on the effectiveness of online social networking as a journalistic tool, some entertainment reporters underlined the challenge of balancing between their private and professional life. They emphasised that as journalists they had to exercise restraint in terms of revealing their own biases and political preferences on their social network profiles. This approach echoes Berger's (2005, p. 1) observation that Southern African journalists are far from "lacking when it comes to critical perspectives with ICTs and global information networks." Journalists also alluded to challenges embedded in independently verifying information sourced from Facebook by highlighting that some people tend make things up or exaggerate issues, especially given the very nature of social networks as social settings.

The Impact of the Internet on Practice and Professionalism

Generational tensions between senior journalists and junior reporters highlighted the impact of the Internet on practices and professionalism. The conservative senior journalists argued that the Internet was taking away the 'human face' of journalism as journalists were now spending more time in the newsrooms wedged on their armchairs, surfing the Internet for story ideas instead of being out in the field directly observing the events on which they report. Junior reporters on the other hand, supported Internet-based journalism. They saw their senior reporters' insistence on 'shoe leather' reporting as counterproductive. One junior reporter at *The Standard* vividly expressed his frustration with his seniors' professional doggedness by suggesting the need to "digitise their minds."

Related to the question of 'armchair journalism' were ethical issues around plagiarism and the failure to verify content from the Internet. Although plagiarism has always been a problem in Zimbabwean newsrooms (Chari, 2009), my observations confirmed that the Internet has made the problem "increasingly simple and tempting" (Pavlik, 2001, p. 47). The narrative presented earlier of my observation of a health reporter writing a story on the decline of HIV/AIDS prevalence in Zimbabwe provides compelling evidence of how plagiarism has become taken for granted as evident in the reporter's oblivious 'copying and pasting' of material from various Web pages to his story. This problem is also exacerbated by the absence of formal policy frameworks and training on the use of the Internet as a journalistic tool.

Overreliance on the Internet also seems to cement established relations between elite forces and newsmakers. As the Glasgow University Media Group (1980, p. 114) argues "[a]ccess is structured and hierarchical to the extent that powerful groups and individuals have privileged and routine entry into news itself and to the manner and means of its production." The appropriation of social networking sites by entertainment reporters discussed earlier provides a good example of the Internet's potential to 'skew' news access through limiting the scope of the stories covered to the interests and agenda set by those with regular access and knowledge of how social networking sites function. As the entertainment editor of the *Sunday News* explained:

> The problem with sourcing stories from social networking sites is that it slants your stories towards a few people who use the platform and therefore leads to journalists missing a lot of good ideas outside the networks. Remember there are many entertainers in this very country who can't tell the difference between the mouse and the keyboard of a computer, let alone, these social networking sites....

This response highlights the social impact of the digital divide, in particular the question of access to technology which includes "(digital) skills or competencies" (van Dijk, 2006, p. 224). Thus, those who actively participate on social networking sites and related platforms become the "primary definers" (Hall et al., 1978, p. 61) of entertainment news and those with limited access are somewhat marginalised especially as journalists contend with the pressure to meet deadlines among other professional pressures.[6] The sociology of journalism has also proved that reporters tend to privilege sources who wield "social hegemony" (Schudson, 2000, p. 184) mainly because they have the appropriate infrastructure to guarantee a reliable and steady supply of raw materials of news production.

The Internet has also facilitated the bourgeoning of 'underground' extra paid work among journalists in the mainstream press. Most journalists clandestinely correspond for foreign news organisations as a way of supplementing their poor salaries as well as surviving the economic challenges facing the country. Although moonlighting has always been part of mainstream journalism practice in Zimbabwe, journalists' exposure to international news organisations through the Internet—coupled with the prohibition of foreign media houses from practising in Zimbabwe[7]—has cultivated a mercenary approach to journalism. However, these opportunities are fraught with negative professional and ethical implications as journalists strive to maximise their financial gains. As one senior editor at *The Zimbabwe Independent* explained:

> We have had instances where we discuss rumours [at the] press club and ten minutes later you find the rumour on the Internet as news. A case in point was during the March 29 [2008] parliamentary and presidential elections, rumours were circulated [at the press club] that Mugabe was losing and that his sister, Sabina, had collapsed and died in shock. In no time at all, there was a story on the Internet that Mugabe's sister had died. Embarrassingly, two hours down the line the government was denying it and up to this day Mugabe's sister is alive!

From the above, it is clear that the speed with which the Internet allows for the publication of stories also fuels the publication of "unsubstantiated and often highly opinionated stories" (Moyo, 2007, p. 91).

Email as a Journalistic Tool: Trends and Patterns of Use

In general, the centrality of email as a newsmaking tool in the newsrooms studied was rooted in its wide-ranging appropriations and merits over traditional modes of communication such as the fax machine and the fixed phone. Evidence from newsroom observations and interviews revealed that the email technology was predominantly used for interviewing sources; receiving news alerts and press releases;

subscribing to different interest group listservs and interacting with readers. I discuss these in turn below.

Email was described as more effective for interviewing high-profile sources in the private sector and non-governmental organisations and corporate organisations than for sources in the public sector and government-affiliated institutions. For this reason, the technology was more strategic for beats whose news coverage routines revolved mostly around the private and corporate sectors such as business reporters. These differences in the use of email find explanation in Fishman's (1980, p. 36) observation that "sources differ as to where they can be located; at what time they are available [and] *by what means they may be tapped*" (emphasis added).

For most journalists, email interviews were strategic as they provided a written record of information exchange between reporters and their sources, hence leaving little room for accusations of misquoting a source. As one reporter at *The Herald* explained: "One good thing about email is that it constitutes a record which you can easily retrieve in the event that your source later alleges that you misquoted them or fabricated issues." For this reason, a number of reporters argued that email interviews could make for more precise reporting as journalists simply copy and paste relevant sections of the email directly into a story. Similarly, journalists pointed out that email gives sources a chance to provide well-thought-out answers rather than top-of-the-head responses that may miss out on critical issues. Using email as an interviewing tool was, however, was largely predicated on prior relationships (cultivated over time) between reporters and sources. This relationship made evaluating information sent through the technology easier for reporters (Reddick & King, 1995).

Email's potential for direct communication was seen as facilitating direct access to public officials and top-level management who are always overscheduled and difficult to get hold of. It was also seen as central for circumventing "the dependence of journalists on gatekeepers and secretaries to get access to key sources" (Berger, 2005, p. 9). In the same way (as shown in the health reporter narrative above), the study also established that email breaks down the barriers created by distance through its capacity to bridge time and space differences between reporters and sources.

A number of journalists also used email to receive daily news alerts from various Internet news sources such as Google News. They also subscribed to listservs administered by various media and human rights organisations in the country such as the Media Institute of Southern Africa and the Crisis in Zimbabwe Coalition. These organisations regularly sent email alerts, which sometimes provided story ideas and leads. Explaining the importance of online material, Reddick and King note that sifting through material churned out on online platforms such as listservs, is like listening to conversations "among people who are deeply interested in a specific subject" (1995, p. 189).

In all the newsrooms email was also used to facilitate what can be referred to as 'dialogical journalism' between journalists and readers. The all-purpose email addresses provided on the print editions of the newspapers allowed readers to communicate directly with the newspapers for various reasons such as responding to stories; voicing concerns on the coverage of specific issues; and providing tip-offs. In addition to the general newsroom emails, some columnists provide their personal email addresses below their columns, thus enabling readers to communicate directly with them.

Ambivalences and Fears in the Use of Email as a Newsmaking Tool

While many journalists glorified the speed, flexibility and efficiency of email, others (mostly the 'old school' journalists), were wary of using the technology as a journalistic tool. The tendency for email to promote laziness and the risk of using unreliable sources were among the biggest concerns raised by senior journalists. They also argued that email interviews eliminated the spontaneity that characterise face-to-face conversations. As one veteran journalist at *The Herald* pointed out: "email is *lifeless*, devoid of *intimacy, pauses, chuckles* and *spontaneous reactions*, all of which add the essential context to a story" (emphasis added). The different conceptions of the value of email as a journalistic tool between senior and junior journalists were perhaps a reflection of the 'generational digital divide' widely perceived in terms of the 'usage gap' between age groups with different social classes and educational backgrounds (van Dijk, 2006).

Journalists' fears and ambivalences in deploying email for journalistic purposes were also rooted in the censorship and security concerns associated with newsroom management's use of spyware to monitor and track email traffic in the newsrooms. One Information Technology manager at *The Herald* confirmed the presence of a system that monitors journalists' email traffic within Zimpapers: "We regularly do random checks for abuse of our online facilities from our main server…if we are not sure of the nature of the email content we simply intercept and quarantine it." Journalists in the private press were equally cautious in their use of email because of fears and suspicions of snooping as was explained by one senior reporter at *The Standard*: "We are aware that management has surreptitiously installed malicious spyware on our computers to monitor emails, so we are very cautious." These fears were also related to the general telecommunications regulation and censorship environment in the country.[8]

Although the censorship environment made it difficult to deploy email as a newsmaking tool, journalists were "far from being mired in 'backwardness' or passively awaiting external salvation" (Berger, 2005, p. 1) in regard to using the technology for journalistic purposes. Thus, where the use of company email is risky, journalists resorted to web-based emails such as Yahoo, Hotmail and Gmail, which they saw as more secure and difficult to monitor.

Conclusion

This chapter has given insights into practices and cultures emerging with the deployment of the Internet and email in the Zimbabwean press. The study generally affirms findings from previous research, precisely that new technologies offer journalists a wide range of resources and technological possibilities to work with (see Pavlik, 2001; Machill & Beiler, 2009). The technologies have spawned flexible work routines as well as enhanced the journalists' control over news production while also leading to reduced time for background research and improvements in meeting deadlines. In this sense, news values and the newsrooms' conception of immediacy in newsmaking processes have assumed new meanings. The study further shows a widening exposure of journalists to news and inevitably a growing participation of citizens in newsmaking even if only those citizens with access and the means to deploy the technologies are involved. This points to a subtle but significant gradual dispersion of the newsrooms' monopoly in defining what constitutes news. In other terms, journalists no longer speak *ex cathedra* as they used to do before the 'online age.'

However, a number of contextual factors have a bearing on the uses of the new technologies resulting in 'local context appropriations' that move beyond a simple substantiation of early studies. Thus, while all the newspapers have an online presence, they still largely duplicate their print editions and have not altered their news production patterns in keeping with the general trends in online journalism such as regularly updating their news content, placing emphasis on breaking news and the immediacy of news coverage. In addition, the Internet censorship environment prevailing in the newsrooms and the country at large, as the study has shown, also impacts on its appropriation by journalists. It has also emerged that the increase in the use of new technologies has opened up a plethora of 'old' ethical and professional challenges with a potential to erode some of the vital and valued practices of the profession such as face-to-face communication in newsgathering.

In conclusion, it is worth noting that the 'localised' appropriations of the Internet and email in Zimbabwean newsrooms reinforce the need to qualify substantially any suggestions of technological determinism. Technology by itself is not a relevant explanatory variable of practices in the newsroom rather, its appropriation is embedded in contextual factors that mediate adoption processes as well as shape and constrain how journalists deploy the technologies.

ENDNOTES

1   The social constructivist approach taken up in this study broadly finds support in the theoretical orientation of the first volume of *Making Online News*, particularly the concern to demonstrate that technologies are, in fact, socially and culturally shaped and that their appropriation is inextricably embedded in the context in which they are deployed (see Domingo, 2008c, pp. 15–28).

2   A common feature of these online newspapers is their "latent opposition slant" (Chari, 2009, p. 11) and their determination to unearth the truth by exposing "corruption and human rights abuses by the Mugabe regime" (Moyo, 2007, p. 84).

3   My research benefited from my prior connections with the research context. Not only was I conducting the research in my 'native' country, but I was also researching among a social group I was intimately attached to through my professional life as a journalism educator in Zimbabwe. More significantly, my activities in social organisations such as the Bulawayo Press Club (where I was on the executive committee until my move to the UK) ensured regular contact with journalists. This 'insider-status' not only helped me to gain some rapport with the journalists but to also avoid mistakes in the highly polarised and sensitive political context of my research.

4   Diary meetings were very brief informal meetings between reporters and their desk editors (often lasting between five and ten minutes) in which reporters submitted story ideas (referred to as 'diary items' in newsroom parlance) to their desk editors for consideration.

5   As an attempt to curb abuse as well as cut costs, most telephone extensions at reporters' workstations in this newsroom were restricted only to receiving and calling local landlines. So, to make long-distance calls, including calling mobile phones, reporters had to use a supervised telephone situated next to the news editor's desk.

6   This argument, however, does not seem to extend to other beats such as political and business reporting as most news providers in these beats appear to have ready access to the appropriate infrastructure that guarantees them access and visibility to newsmakers.

7   At the time of doing fieldwork for the present study, many foreign news organisations deemed hostile to government policies had been banned from practising journalism in Zimbabwe.

8   In 2007, the government promulgated the Interception of Communication Act which provides for the setting up of an interception centre to listen into telephone conversations, open mail and intercept emails and faxes. The Act also compels Internet service providers to install equipment to facilitate full-time monitoring and interception of communications.

PART TWO

News Redefined

CHAPTER SIX

Routinizing Breaking News
Categories and Hierarchies in Danish
Online Newsrooms

Jannie Møller Hartley

This chapter revisits seminal theoretical categorizations of news proposed three decades earlier by US sociologist Gaye Tuchman. By exploring the definition of "breaking news" in the contemporary online newsrooms of three Danish news organisations, the author offers us a long overdue re-theorization of journalistic practice in the online context and helpfully explores well-evidenced limitations to online news production, such as the relationship between original reporting and the use of "shovelware."

EDITORS' NOTE

"My list is quite short today—most of the day will be dedicated to our new Prime Minister. However, we do need some business stories as well, and I guess that we will also get some soft stories from Cannes [Film Festival] and perhaps something glamorous. Let's keep an eye on Michelle Obama."

This editorial meeting in the newsroom of the online newspaper *Politiken.dk* was special. The night before, the Danish Prime Minister had resigned and *Politiken. dk* had been the first to publish the story. They were ahead, in breaking-news mode and felt euphoric. Several online journalists were given the special task of following the story as it developed during the following 24 hours.

Using empirical data from three Danish newsrooms this chapter aims to clarify the much-used concept in the world of online journalistic practice: *Breaking news.* What happens when online journalists are suddenly confronted with a non-routine occurrence and how we can understand how these routines in a new media context are different than how newswork was described by Gaye Tuchman (1978) in her seminal analysis more than 30 years ago?

The quote from the editorial meeting and the decisions taken to organize the day in the newsroom remind us that online journalists still use categories in order to "routinize the unexpected"—even when the staff is dealing with highly unexpected and unusual stories. Tuchman showed how newsworthiness is negotiated in the newsroom, with both reporters and editors sharing an underlying knowledge of not only the different categories of news but also how the covering of the different categories should take place (Tuchman, 1978, pp. 31–33). The aim of this chapter is to broaden the discussion of routines on newswork by examining the concept of *breaking news*—a term that only became populated some time after Tuchman's study was carried out—thus revisiting Tuchman's typology of news and examining its analytical power in a new media context.

Research has shown how, since an online publication does not consist of a specific number of pages, has no single deadline, and can be constantly re-edited, there is a pressure for immediacy in the news production process (Karlsson, 2007; Lindholm, 2008; Domingo, 2008a). Following this, it can be argued that the majority of the content online can be characterised as Tuchman's *developing news*— the covering of news as the facts are unfolding (Tuchman, 1973, p. 115). But empirical results also suggest that online news production is not actually continuous but subjected to soft deadlines and different immediacy criteria for different stories (Domingo, 2006). Hence we must ask, if new platforms and new ways of organising the processing and publishing of content online focusing on immediacy lead to new categories or if old categories are re-negotiated in the new context?

These negotiations present when typifying a *breaking news story* are examined by applying Pierre Bourdieu's field theory—thereby bringing values of newsworthiness back into the analysis of routine newswork. Thus the second aim of the chapter is to make it explicit how online journalists are using the categorisations to challenge or support hierarchies within the journalistic field in order to position themselves as specialists in Tuchman's *developing news*, aiming and striving for *breaking news* and the "exclusive scoop," as the trademark of online journalism in a competitive media environment where immediacy rules (Domingo, 2008a). The final part of the chapter sums up and offers a model of the interconnected factors influencing the newsroom staff's typification[1] of *breaking news.*

Tuchman's Typology of News

Tuchman distinguishes a first level between *hard news* and *soft news*. The main difference here is, that *soft news* is characterised by the fact that the story can be published any day and is often a feature or a human interest story, whereas *hard news* is either based on a sudden unexpected event or pre-scheduled by, for example, agencies or people outside the newsroom. *Hard news* is divided into 3 subcategories, where *spot news* is the unplanned event that has to be covered straight away, a characteristic shared with the category *developing news*. But *developing news* is, according to Tuchman, described as "unscheduled coverage of news as facts are unfolding" and as often viewed differently by television staff compared to newspaper staff because of the different time-modes involved in television and in newspaper production. The fact that television potentially can interrupt the day with an important piece of news makes them classify it as *developing news*, whereas the technology involved in making the newspaper limits the amount of deadlines and thus they see the event as *spot news*. *Spot news* thus unfolds in a short time period and the basic frames and boundaries of a story can be quickly decided, while *developing news* takes longer to be framed although the facts remain almost predictable.

Continuing news is characterised as a series of stories on the same subject, prescheduled in the sense that the coverage can be planned ahead (Tuchman, 1973, p. 123). The categories and their relation to the different factors shaping the production of news— time, technology and planning—are shown in Table 1.

TABLE 1: News typifications.

		How is an event scheduled?	Is dissemination urgent?	Does technology affect perception?	Are future predictions facilitated?
Soft News		Non-scheduled	No	No	Yes
Hard News		Unscheduled and prescheduled	Sometimes	Sometimes	Sometimes
	Spot news	Unscheduled	Yes	No	No
	Developing news	Unscheduled	Yes	Yes	No
	Continuing news	Prescheduled	Yes	No	Yes

Source: Tuchman (1973: p. 117)

Tuchman provides a couple of examples of what happens when the newsroom staff is confronted with a highly unexpected event and she calls these instances *"What a story!"* Through examples Tuchman did operationally define this type of story, however, she did not fully develop the concept. While Tuchman's categories are useful when studying online news production, the descriptive and organisational approach also has shortcomings. In the editorial meeting referred to above there was a clear differentiation between the categories—some seem more important and some were mere "filling the site." The problem with Tuchman's typology is in other words that it is "flat"—the typology offers us a way of distinguishing between different types of news stories, but it does not show how some are valued more than others and thus how the categorisations are negotiated inside the newsroom and between different news sites.[2] For this I found the critical, relational and historical approach of Pierre Bourdieu's field theory useful.

Categorising as Positioning Strategies: A Field Theory Perspective

Using field theory means seeing journalistic practice as an ongoing struggle between different journalistic positions within a social space called a journalistic field (Bourdieu, 1998 [1996]). Space does not permit me to discuss the theory fully here nor describe its more controversial suggestions, but I will highlight why some of the concepts were useful when examining news categorisations in online newsrooms. The relational aspect (Bourdieu, 1993, 1998, 2005) is apparent in Bourdieu's conceptualisation of fields as hierarchical social spaces, where each position in the space is defined by its relation to other positions in the space (Schultz, 2006). As Bourdieu notes, categories are socially constructed and at the same time acquired socially (Bourdieu, 2005: 36). The very categorising in a field perspective becomes a sort of positioning in the struggle for capital (recognition) and thus power in, for example, the journalistic field. Applying a field theory perspective on an ethnographic analysis of Danish journalists, Ida Willig (Schultz, 2006) argues that the hunt for the *exclusive story* is an essential part of journalistic practice because the category is recognised as particularly valuable within the journalistic field—with a story "planned," "found" and researched by a specific media organisation, the newsroom can position itself in a privileged spot over its competitors. *Exclusives* are thus in the Danish journalistic field related to high status and potentially increase the symbolic capital of the media organisation that originated the story. This is provided, of course, that other media organisations follow up by quoting and thus signalling that this is a "good" story.

Thus, the field perspective supplies the analysis with relational categories such as *the exclusive, the quote story, the wire story*—categories that bring power and relations between the different online news sites and platforms inside the media organisation into the analysis—alongside Tuchman's categories of more organisational character.

Sample and Methodology

The range of candidates for the case studies for ethnographic observation was chosen on the basis of the initial mapping of online news publications and on the basis of a content analysis of eight Danish online news sites carried out in November 2008. The content analysis showed that the online publications differed along a continuum, where the online platforms in one end were characterized by a *decentralized* production of news, leaving the often quite small online desk with the task of distributing content from other media platforms in the organization, from wire agencies and from other news sites (that is, "shovelware"). On the other end we could find a more *centralized* model, where much more online content was produced by the online staff for the online desk only. It is important to notice that none of the news sites was online-only,[3] and none had more than 45 percent of content produced by the online news staff for the online platform only. The cases were then chosen from each end of the continuum and for different attributes as shown in Table 2.

TABLE 2: Overview of cases

	Politiken.dk	Dr.dk	Nordjyske.dk
Media Model	Print/Online	Radio/Television/ Online/24 Hour News TV	Print/Radio/ Online /24 Hour News TV
Business Model	Non-profit foundation	Public Service	Commercial
Scope	National	National/Regional	Regional
Convergence Model	45% online production: most centralised	18% online production: both de-centralized and central production	8% online production: most decentralized
Size of Online Desk	45, including the technical staff	30, including technical staff	11, excluding technical staff

The cases are then selected for their diversity, which represents a "most-different" approach in case study research (Andersen, 1997, p. 107). I was present in each of the online newsrooms in periods of eight weeks from May to October 2009, mostly during daytime but also during nights and weekend-shifts. I was present at editorial meetings and in the daily production, sometimes sitting next to one of the journalists and other times next to the editor. During my stays I carried our 35 in-dept interviews lasting an average of 45 minutes with journalists and editors from both online platforms and other media platforms.

The Case of Casper

A reporter from another department calls the front-page editor of the online site telling him that there are police cars at the harbour. The source on the phone happened to be walking along the fjord and had already alerted police that there is a body floating in the harbour. The front-page editor runs to the news editor and they quickly discuss and agree on a plan. The only online journalist in the newsroom at the time is just finishing his shift but nevertheless agrees to make the first call to the police.

We are in the newsroom of *Nordjyske.dk*, consisting of a set of tables placed in a large circle in the middle of the larger newsroom. The general news editor, the editor of the online site, the radio reporter, and the television reporter are placed in front of two computer screens each, with their backs to each other, facing the reporters from the newspaper sitting in the room around them. Talk between them intensifies.

The online reporter reaches the chief of police. The reporter's email is down, so he delivers what the police said face to face to the homepage editor, who writes the story immediately and posts it online. They find a journalist from the newspaper who can drive to the harbour. He makes a second call to the police before heading to the harbour. The atmosphere is frantic. Everyone helps. The online site editor shouts that he needs the exact day of the disappearance of the teenager who has been missing for nearly a week, and he soon gets an answer. He puts this in the story, even though it has not yet been confirmed that the body belongs to the 16-year-old, Casper, whose disappearance had been front page news for days.

After publishing the first version, he puts more background information at the bottom of the story. The radio journalist sitting next to him notes that a competing online newspaper, *TV2.dk*, also has the story now. The homepage editor calls the mobile phone of the reporter who is on his way out the door. *EB.dk*, another competing online news site, has a story that confirms the body as Casper's. "Did the police said something to you about this?" he asks, and underlines that the reporter must ask in the interview if the police can confirm in any way that it is Casper.

A journalist from another editorial desk calls to say that the regional competing TV channel *TV2 NORD* has a story that claims that Casper's keys were found on the corpse. "It's probably what we will get from our own reporter in a bit, but we want it now, so we'll have to quote *TV2 NORD*," the editor replies. He creates a new article and chooses the same headline as *TV2 NORD*. An hour later the journalist has interviewed the police on the harbour and phones the editor with the quotes. But the homepage editor decides not to write a new story, as the reporter could not get the information about the keys confirmed. Thus he cannot avoid quoting the competitor but decides that he will make the journalist call the police again when he gets back to get a final confirmation on the keys.

The news editor is also annoyed that *Nordjyske* doesn't have the story that Casper's keys and mobile have been found. "We have quoted *TV2 NORD* for hours now and it's not ok," he says. He cannot find the homepage editor who has gone for a smoke. "It will soon appear everywhere, so that we really should get it confirmed," he says to himself as he goes to find the homepage editor.

Breaking News vs Tuchman's Typology

The events above happened during my stay in the newsroom of *Nordjyske.dk*—a 16-year-old boy went missing after a school party. The story of Casper was described many times by journalists and editors as a typical example of a *breaking news story* and an ideal story for the online desk, because the story was unfolding and they were covering the case "in the moment"—a feature Tuchman ascribes to her category *developing news* (Tuchman, 1978, p. 55). If we use her typology to examine the category of *breaking news* the empirical data shows that dissemination is always urgent and the news coverage is hard to plan—in the case of Casper it became a sort of desperate struggle to get the story out quickly as the facts were unfolding. Like *spot news* and *developing news, breaking news* is also often un-scheduled—it cannot be planned ahead. That leaves us with the question of technology. *Breaking news* was in the empirical data, because of the, at least potentially, continued deadlines, often characterised as *developing*. The interviewed journalists used the two terms interchangeably. It is even implicit in the word "breaking," as the ing-form indicates that the coverage of a certain event is ongoing (the same indication is given by Tuchman in the two categories *developing* and *continuing*). Thus *breaking news* shares most traits with *developing news* and the urgency with *spot news* and the "covering of several stories on the same subject" with *continuing news*.

However, a *developing* story seems in Tuchman's case to refer to processes inside the newsroom during the day and thereby processes hidden from the audience, whereas the coverage of *breaking news* in an online newsroom is happening continuously, compressing the time between the gathering of facts and the distribution of them in the form of news. But this is not the only difference.

The editorial meetings was often concerned with defining which 3 focus stories the staff could cover as *developing*, which involved calling news sources, finding extra material and new angles during the day. On the other hand it was never discussed which stories that could potentially *break*. And the fact that an event could be described "in the moment" did not always turn on the "breaking-news signs" (the online publication had a special feature, allowing for the top story to be portrayed in a bigger format or indicated with big or flashing letters at the top that this was *breaking news*). At *Politiken.dk* and at *Dr.dk* breaking news could also be identified, because it meant that they had to send an SMS to readers who had signed up for this service.

Furthermore, not all *breaking news* was *developing news*. When asked the journalists would mention a plane crash, a fire or an unexpected death as events that could qualify as *breaking news*. When I was present at the newsroom of *DR.dk* a famous politician and an actress died, and both times the online publication went into "*breaking-news mode.*" One could argue, that when someone dies, there is no development, and in Tuchman's typology this story would be *spot news*. The facts were simply there. But firstly the way the journalists reacted and dealt with these stories made it into *developing news*. They published comments from other politicians, from the readers, old movie clips, historical articles, timelines and in that way they automatically created a development in a story as they kept posting additional facts and angles to the story in an accumulative manner. Secondly, when asked why the story of the deceased actress was also *breaking news,* a journalist said "because it was really unexpected and a story with an interest for many people." This factor seems related to the newsworthiness and audience rather than specific organisational tasks and is in some ways similar to Tuchman's category "*what a story!*" (Tuchman, 1973, p. 128). Like in the opening example, the atmosphere in the newsroom was euphoric when there was a case of *breaking news*. The reporters also rubbed their hands and talked over lunch about the "great news day" and told me how lucky I was to be present on such a fine news day. In this light *breaking news* can be seen as a combination of any of Tuchman's *hard news* categories and the type of story that brings about a "*what a story!*": the story must be highly unexpected and of high relevance to the readers.

My observations showed that because both *breaking* and *developing news* were identified as the ideal task for the online desks, the journalists would try and make *continuing news*—into *developing news* and thus be prepared if the story would break. However *continuing news* could only break if something really unexpected happened. A trial is a good example of *continuing news* in Tuchman's typology, as the trial is pre-scheduled and thus planned, and yet it is still necessary to get the results of the trial published as quickly as possible. However, my observations showed, that despite the fact that a trial had finished they would continue writing on the series of stories, by calling new sources, establishing chats with the users, or asking experts to provide analysis. This was often described as "keeping the pan boiling" or writing a "filler story," and this was often covered using wire-service stories in between doing their own stories. As this online journalist explains:

> We do not investigate the municipal reform or something like that on the online desk—but we put ourselves in the slipstream of it and sort of write some of the leftovers. And as purely personal or journalistic criteria of success we basically need to react fast so we need it to be on the spot where it happens. We have, for example, also started to write—when we have these kind of cases that are ongoing—the arrest, preliminary hearing, court hearings…it could be several months in between this myriad of little events—small stories and follow-ups—we make some—others we take from the wire, and they are organized together in a long list.

In this quote we see that re-working *continuing news* to a kind of *developing news* enables them be the first ones to bring the outcome of an event. This becomes the journalistic criterion of "good journalism" because they otherwise only quote others and distribute material produced elsewhere and thus indicates the importance of the *exclusive story* in the online journalistic practice. This leads to writing the story even before the event has taken place. If they were unsure of the outcome of the trial, they would just have to change the headline and the beginning of the article. As this interviewed journalist explains:

> What we have begun to do is that the day of the ruling, if we know that a certain trial will end at 12, we write everything together and then we just copy and paste a quote into the story when it arrives. This means that we can have the entire case summed up, when the conviction falls—immediately.…

Many of the journalists could even mention at least one case where an unfinished story had been published by mistake, and technology seems here to play an even bigger role than in Tuchman's study as the archive function enables the newsroom staff to publish the story set to a timer.

What is clear from these examples is both that Tuchman's typology is still very relevant in a contemporary digital context and that all of Tuchman's categories can potentially become *breaking news:* the way they deal with these unexpected events is to routinize them by adapting strategies from everyday work routines. But although online journalists especially value *breaking news*, the interesting thing is that these stories seem very rare in practice. When interviewed, journalists would normally refer to 2 or 3 stories like this one over the past year—stories where the journalist felt that this was "good online journalism." Thus *breaking news* is a specific kind of news category—one could say "the cream" of news. But in order to understand this we have to look at how they collaborate and compete with other platforms and media organisations.

Surveillance and Permission to Produce

A big news day would be different from a "normal" day in that *Dr.dk* and *Nordjyske. dk* would collaborate with the other platforms and work in teams with television and radio staff. But the most apparent difference was the fact that a *breaking story* would give the online desk the right to publish stories without asking whether this was a story being covered elsewhere in the organisation. *Breaking-news mode* made it their story and gave the online desk the possibility of moving from *distributing* news to *producing* news. This is an important point. The fact that the journalists are striving to be self-producing—and thus *breaking-news mode* giving them this opportunity—is likely to create an increased focus on *breaking news* and news where the online staff could argue that other media organisations would follow, which meant they could not keep the story from being published. As an example, an editor asked after hearing about a story from a reporter "do we have it solo?" When the reporter said that he had seen several other journalists on the spot, the editor answered: "Then run it quickly." In one newsroom it was specified in a manual: "stories must be published online first, especially in the case of *breaking news* where we can expect that other media organisations also have the story, or when we hear from sources who might also have contacted other media organisations." In the example of the court trial above, the journalists were most of the time distributing "leftovers" and "lying in the slipstream," but when they were among the first with the result of the trail, they felt like "real" journalists, *producing* news ahead of their competitors.

This again shows how the internal hierarchies influence how the news workers categorize the stories. In two of the newsrooms the online staff often had to check with the other platforms to determine whether they were covering the same story. The argument was that if the television people were covering it, the online platform could just write a version afterwards, but the other way around was no good for the television journalists, as they would be lacking the moving pictures. There were only two ways that the online staff could work around these hierarchies—when covering *breaking news* and when they had an *exclusive hard news* story (often involving some investigative reporting). Here there was a clear difference between the online newsroom with a more centralised production—in this case *Politiken. dk*—and the other two organisations with the production for the online platform coming from other desks (television, print, business desk, etc). *Politiken.dk* with the most centralised production, more staff to produce for the Web and less distribution from the newspaper, the staff did not have to coordinate the coverage of their stories with colleagues from the print newspaper.

Breaking news also meant that they could email their *breaking* story (after checking if anyone else also has the same story) to the news agency that will then distribute it to all the other media organisations. This meant that they could see their story quoted on other news sites, which would be accompanied by talk among the reporters: "Have you seen that our story is quoted on *Berlingske.dk*?" Thus by quoting stories they credit the quoted news site, and by being quoted they confirm their own judgement, when initially pronouncing "what a story!"—increasing their symbolic capital in the field. In the Casper case above, the reporters and editors struggled to "hold on" to the story, and the difficulties getting facts confirmed meant that they lost their lead and thus the ability to go *breaking*. Thus whether it is categorised as *breaking* also depends on what other news sites are doing—the more coverage, the bigger it seemed to be. And yet, if too much time has passed, e.g., they discover the story or the event too late, it will just be covered as a *spot news story*. It was also primarily the news organisation that heard of the news first (and thus has it as an *exclusive*), the one able to label the story as *breaking news*— when the story has already been published at several other media it is no longer considered *breaking news* but merely *developing news* or *spot news*—unless the newsroom is among the first ones to follow.

Thus, breaking news is also a way of giving credit or discrediting the competitors. This is important. In a field theory lens we can see this striving to be ahead of competing news sites and the hunt for the exclusive through the massive focus on *breaking news* as a form of journalistic capital and the existing known forms of capital, namely, that of *production,* in the field. Not surprisingly the constant deadline and re-editing of news online and the large amount of shovelware increase the focus on the type of news particularly suited for this process. However, the examples above show that this issue might not only be related to technology but that time and relational positions in the journalistic field are equally important. Early in the Casper case where they were ahead of competitors, they were eager to call it *breaking news* and followed closely how they were quoted by competing news sites. And when they "lost" the story to the competing regional television channel, they despaired—they did not feel they could categorize the story as *breaking news*.

Conclusion

Comparing categories used by Danish online journalists today with those made by American journalists more than 30 years ago enables us to see that despite technological innovations many typifications remain the same. However when looking at online news production, Tuchman's typology needs to be somewhat expanded to account for new categories in a fast media environment enabled by new technology. By looking at the values that online journalists attach to the different categories during newsroom observation this chapter has argued, that *breaking news* can be seen as "the exclusives" of online news production. When they manage to be the first to publish a certain news story by having continuous deadlines, they can "force" other news publications to quote them, allowing the news site to build symbolic capital in relation to competing news publications. The other important factor is, whether the story is of great interest to the audience, which can compensate for a potential lack of exclusiveness.

To summarize, breaking news is influenced by the interconnected factors of audience appeal and exclusiveness. Tuchman's factors of time, technology and planning are related to both of these, e.g., the news website could lose exclusiveness if too much time has passed; the coverage of *breaking news* could be routinized and technology played a part in this. And equally the audience appeal is related to the news agenda at the specific time of the event. However the two factors need to be present at the same time in order for a story to break, but a news story can be more or less exclusive and have more or less audience appeal as shown in the continuum grid below (Figure 1).

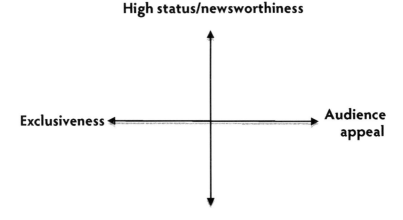

High status/newsworthiness

Exclusiveness ← | → **Audience appeal**

Low status/low newsworthiness

FIGURE I: Factors influencing breaking news

However the question is whether these are new for online production? Literature on production of news since Galtung and Ruge's study (1965) on news values has talked about identification, relevance and unexpectedness, and Willig's study (2006) highlights the importance of the *exclusive solo story*. This analysis of the routines of *breaking news* shows the interconnectedness of these factors—hence hard to achieve in practice but very much present as an ideal in online journalistic production.

Furthermore, there were remarkable differences between the three newsrooms that were chosen because of their different percentages of original production and shovelware. Online newsrooms with a high amount of shovelware were less able to cover news as *breaking stories* because of their dependence on the production from other platforms in the media organisation and wire stories. With a more central production *Politiken.dk* were able to increase their capital in the field by making *spot news, developing news* and *continuing news* into *breaking news* and by at the same time having fewer *wire* and *quote stories* when such coverage of an event took place.

In future research we need to look at not only how events are covered and newswork organised according to time, planning and technology but also how these events are covered in relation to each other and how applying the field perspective allows us to see how categorising news stories is related to positioning strategies on the part of the online publications towards and in relation to other publications (both online and off-line). It shows that categories are not only "routinising the unexpected," they are also constantly dividing, supporting and challenging hierarchies in the journalistic field—enhancing our understanding of why immediacy rules in most online newsrooms.

ENDNOTES

1 Tuchman suggests the use of typification, meaning classifications where their meanings are constituted in the situations of their use and notes that the use of classifications alone is inadequate (Tuchman, 1973, p. 112)

2 Another problem with Tuchman's organisational approach is that the typology is based on news as events—thus leaving no room for analysis of, for example, investigative news.

3 At the time the content analysis was carried out (2008) only very small niche online-only news sites existed in Denmark and they were not considered for the study.

CHAPTER SEVEN

Making Online Features
How the Discursive Practice of an Online Newsroom
Affects Genre Development

Steen Steensen

The Nordic countries have emerged as leaders in contemporary news production research as this chapter and others in this volume attest. In this case, the
author advances a more comprehensive analysis of online news production
through his exploration of the often-neglected genre of feature journalism.
Through an ethnography at an online newspaper in Norway and content analysis of the features produced by the newsroom, the discussion contributes fruitful theoretical interpretations on the development of online news formats as
stemming out of the newsroom practices and the tensions between online and
print cultures.

EDITORS' NOTE

How are genres actually formed? This question is becoming increasingly relevant
because we live in an age in which means of communication are more numerous
and complicated than ever before. How texts appear in the news media is more
dependent on context than ever before, simply because the news media are used in
many more contexts than previously, for many more purposes on a greater variety
of technological platforms. This means that new genres arise in close interaction
with social contexts. Without understanding the social context of a text that appears

in the media it is therefore difficult to understand the text's purpose and function, i.e., what type of genre it constitutes.

The aim of this chapter is to show how a methodology that combines textual analysis with observations of the discursive practice of which the texts are part is better suited to understanding genre development—and thus the changing social function of journalism in new media—than textual analysis or ethnographic enquiries alone. Studies of news production in general and online news production in particular tend to focus on other aspects of production than the actual text, like how the role of journalists develops in new media (Domingo, 2006a; Domingo et al., 2008; Nygren, 2008), how new technology diffuses in online newsrooms (Boczkowski, 2004a; Küng, 2008), and how convergence affects news production (Erdal, 2008; Klinenberg, 2005; Singer, 2004). Studies that combine genre awareness with newsroom observation remain scarce. An exception is Boczkowski's study of the differences between soft and hard news production in an Argentinean online newsroom (Boczkowski, 2009), but even in this study the texts themselves are left out as research objects.

This chapter draws upon empirical data from a case study of the Norwegian online newspaper *dagbladet.no* and the way in which it has developed a separate section for feature journalism. The investigation is framed by the research question: How is the development of feature journalism as genre influenced by the discursive practices of an online newspaper?

Genre, Discourse and Discursive Practice

The research is framed by a pragmatic understanding of genre based on Carolyn Miller's often-cited 1994 article "Genre as Social Action" (1994). Miller emphasises the need to study the situations in which texts appear in order to understand what genres are. Berkenkotter & Huckin (1993) take Miller's understanding of genres a step further and believe that the perspectives of the producer of the text and other production conditions must be studied from the inside if genres are to be understood fully. Greg Philo (2007) introduces a similar view and claims that studies of media texts require studies of the social structures that form the basis for text production. Such an understanding of genres makes it fruitful to view genres in connection with *discourse* and *discursive practice*. Teun A. v. Dijk (1997) names twelve different approaches to discourse analysis, one of which concerns understanding discourse as *strategy*. Such a discourse analysis will seek to expose which strategies are behind the communication, i.e., what the aim of the communicative activity is. In order to understand genres as social action, it is fruitful to search for what the aims of the communicative activity are, both for the producer of the utterance in question and the recipient. Which discourses are represented in a genre tell us something about the genre's social function. The genre can therefore be understood as the sum of the discourses represented in the text and the text norms that are used to realise these discourses.

Discourses are thus represented in texts but originate from outside the text, more precisely in the discursive practice, which the texts are part of, i.e., conditions linked to the production, distribution and consumption of the text (Fairclough, 1995). However, there is no tradition of obtaining empirical information from outside the text in either rhetoric or linguistics. A core aim of this chapter is therefore to identify a methodology which is suited to finding out *how* discursive practice sets the premises for and influences the development of a genre.

What Is Feature Journalism?

The empirical material on which this chapter is based concerns the development of feature journalism in an online newspaper. Feature journalism is most commonly associated with newspaper weekend sections and magazines that feast on celebrities, personal columnists and reportage journalism. I understand feature journalism as a kind of journalistic practice that serves a particular communicative purpose and responds to a specific social need, or "exigence," as Miller (1994) labels it. In Steensen (forthcoming) I argue that this social need, the exigence of feature journalism, traditionally has been to entertain and connect people on a mainly emotional level by accounts of personal experiences of perceived public value. This exigence is addressed by utilising different rhetorical vehicles or text norms. The feature reportage typically embeds a narrative structure, first-person accounts of events and a "colourful" style of writing. The profile interview typically contains in-depth questions and answers between the journalist and the interviewee, colourful writing, personal characterisations and a profile picture or caricature drawing. The personal column is typically characterised by a given length and placement within the newspaper, a first-person narrator, and colourful, anecdotal writing about everyday and current events.

Feature journalism is therefore best understood as a *family of genres* that share a common exigence, which is expressed in the different genres by different text norms. The exigence of traditional feature journalism is expressed by three discourses: A literary discourse and discourses of intimacy and adventure. A (print) feature journalist has therefore traditionally been someone with literary ambitions who seeks to provide the reader with adventures and intimate encounters with people and milieus. The feature journalist might also be intimate with her audience by writing in a subjective style, sometimes even using the personal noun "I."

The question to be addressed in this chapter concerns how feature journalism changes when it moves from a hard-copy newspaper to an online newspaper. I have elsewhere (Steensen, 2009a) demonstrated how the discourses of feature journalism collide with discourses of online communication, in which rapid news reporting, debate and reader involvement are central. The aim here is to find out how the discursive practice in an online newsroom influences and possibly transforms feature journalism as a genre. In order to get closer to answering this, three secondary questions identify themselves:

- What sort of understanding of feature journalism do online journalists have?
- How do texts that are genre-labelled as feature journalism on an online newspaper relate to the traditional understanding of the genre?
- What characterises the social context in an online-based feature editorial operation?

Methodological Approach

I have argued that genre research would benefit from an increased focus on studies of texts' social contexts. When studying genre development in an online newspaper, it is therefore a logical choice to combine textual analysis with observations of conditions linked to text production, distribution, and perhaps also text consumption. At the same time, online journalism and web-based genres in general develop quickly. To capture this change and indicate possible directions of movement for genre development, it is therefore also necessary to construct a longitudinal perspective.

Such a research design generates extensive empirical work and is time consuming. It is therefore difficult to apply such a methodology to several case studies, and this chapter is therefore based on empirical data from an isolated case study of the Norwegian online newspaper *dagbladet.no*. In 2002, *dagbladet.no* became the first online newspaper in Scandinavia to set up a dedicated section for feature journalism. The development of this section forms the empirical starting point of this article. The empirical material generated by this study consists of interviews, document analysis, textual analysis and observation of production processes and production conditions.[1]

Interviews and document analysis. Fourteen journalists and editors on *dagbladet.no* have been interviewed about their understanding of, and opinion on feature journalism and online journalism in general, and the hard-copy edition of *Magasinet* and *Magasinet online* ("Magasinet på nett") in particular. In addition, six other staffers were interviewed for background information on the early history of the section. Together with document analysis of project reports on *Magasinet online* and the journalistic programme for the section, these interviews provide answers about what kind of understanding of genre forms the basis of the section, what the text producers' *strategy* is concerning what and how the section communicates, and how this strategy has developed. The interviews were conducted as semi-structured interviews in which the topic was disclosed in advance. The interviews took place with individual variations, depending on what the interviewees focused on.

Content analysis. In order to find answers to how this strategy corresponds with the actual text production, I have analysed a selection of texts in two limited periods: a selection of 30 articles from 2005, before the section was redesigned, and a corresponding selection from 2009, after the section was redesigned. All the 60 articles were written exclusively for *Magasinet online* and were not printed in the hard-copy edition of *Magasinet.*

The content analysis has four aims: First, I wish to see the extent to which the textual norms from feature journalism as it has traditionally been understood are present in the texts. Second, I wish to see to what extent the text producers' strategies can be found in the texts. Third, I wish to look for online-discursive features of the texts. And fourth, I wish to see if the texts change over time.

The articles have been analysed using a set of codes deduced from the textual norms of feature journalism presented above and adjusted according to the findings of a qualitative discourse and genre analysis of an individual *Magasinet online* story (Steensen, 2009a). I have also based them on the three dominant discourses of the genre: adventure, intimacy and literary. In addition, the articles have been coded to find online-discursive features such as interactivity and the presence of multi-media content. The codes represent a mixture of quantitative registration and qualitative assessment.

Observation. I have been a participating observer at *dagbladet.no* for six weeks in total, in four different periods between May 2005 and November 2007. The aim of this was to observe discursive practice, i.e., how the social context influenced the production process and how the interaction among text production, text presentation and text consumption developed. I asked questions and actively attempted to gain insight into what happened in the newsroom and why it happened. I also sat still and observed the editorial team at work for long periods. This makes the method vulnerable to my own human abilities to make contact with sources and gain their trust. The empirical material that the participating observation generated was written down in a field diary.

Feature Journalism on *Dagbladet.no*

When *Magasinet online* was launched in February 2002, the primary strategy was to create an online arena for readers of the hard-copy edition of *Magasinet* by using Internet tools such as a discussion forums and net-meetings.[2] As the years passed, *Magasinet online* broke away from the hard-copy *Magasinet* and focussed instead on self-produced feature journalism (Steensen, 2009b). Up until 2006 there was only one journalist producing stories for the section full-time, but many other journalists in the editorial team regularly contributed stories. In 2007 the section was changed and expanded according to a plan following an extensive planning period. The intention was to define what kind of journalism *Magasinet online* should contain and to expand its focus.

In February 2009 there were three journalists affiliated to the section full-time. In addition, other staff reporters as well as freelancers and additional external contributors regularly contributed material.

Perceptions of Feature Journalism

The journalists on *dagbladet.no* had a very clear and unanimous and quite conventional understanding of what feature journalism is. Concepts and formulations such as "tell stories," "good writing" and "personal/subjective" are repeated in the interviews. But an interesting feature appears if we distinguish those who had previously worked on newspapers (eight out of 14 interviewees) from those who had only worked online. Half of the journalists in the first group also had a journalism degree, something which none of those who had only worked online had.

The online journalists with a background in newspapers, and particularly those who also had a journalism degree, stood out to a much greater extent as *traditionalists* in their understanding of feature journalism. They were also more inclined to perceive feature journalism as something timeless, a form of journalism that does not relate to the news flow. The journalists without a newspaper background and qualification associated feature journalism with the news flow to a much greater extent. They believed that feature journalism goes "behind the news," "gives background to the news flow" or similar expressions.

However, the two groups had one thing in common: Very few of the journalists referred to aspects of feature journalism that indicated the presence of the discourse of adventure. This contrasts with the view of feature journalism of the editor of the hard-copy *Magasinet*. She emphasised *reportage*—the journalist going out to look up people and milieus—as the central feature of feature journalism.

So how did the journalists perceive the *Magasinet online* section? Everyone whom I interviewed was unanimously positive. The journalists emphasised that the section gave *dagbladet.no* a degree of diversity that no other online newspaper had, that it gave them the opportunity to work with an alternative and different type of journalism online—a type of journalism in which they could take their time and write in a different way than they otherwise did. They were proud of the section and all the journalists stated that they wanted to write for it. In other words, the section had high status within the newsroom.

The question is therefore whether this view of feature journalism and of the section is reflected in the texts which are produced. Which of the two groups' views of feature journalism dominates the texts? Is the discourse of adventure marginalised?

Content Analysis

In 2005, the section had the same appearance as it had been given when it started in 2002. Design elements such as different fonts, the use of colour and the set-up of headlines, lead texts and by-lines had all been borrowed from the hard-copy *Magasinet,* and the section therefore gave the clear impression that it was an online version of the hard-copy *Magasinet*. The design signalled that this section was something other than a place for breaking news—here you would find more dateless material suitable for immersing yourself in.

In the main the articles consisted of long, linear texts illustrated with a series of photographs. The articles in the selection analysed have an average length of 1,268 words, which is considerably longer than what is found to be normal for online journalism in Scandinavia in general (Engebretsen, 2006). The length corresponds to the length of a standard article/reportage over six pages in the hard-copy *Magasinet*. On average, each article was illustrated with eight photographs, which is also much more than usual in online journalism in general, but more or less what is usual in the hard-copy *Magasinet* for stories of corresponding length. But more or less all the photographs (93 per cent) were either provided by wire services or retrieved from the archives of *Dagbladet*. Furthermore, the texts were characterised by these features:

- Two thirds of all the articles could be categorised as news or related to the news flow.
- The articles were very rich in sources and on average used 11.3 sources. 81 per cent of these sources were other websites or articles from *dagbladet.no*'s own archives. Only 14 per cent of the sources were direct interviews con-

ducted by the journalists, and in only four of the 30 articles were the journalist's own observations a source in their own right.

- The articles were equipped with an average of 13.6 hyperlinks.
- Three quarters of the articles were dominated by narrative composition.
- Subjective descriptions and characteristics provided by the journalists were present in only two of the 30 articles. Four of the articles bore the "human interest" imprint, implying that they portrayed personal dimensions of sources.
- Three of the articles had multimedia content (two articles had graphics, one had video as well as photographs).
- Only six of the articles had interactive elements in addition to hyperlinks (two articles were net meetings; three had a leave-a-comment option, and one had a blog linked to it).

The typical article from this period was a long, news-driven narrative that had a lot of hyperlinks and was rich in sources but with few first-hand sources. It was matter of fact rather than subjective and personal. And it used online tools such as multimedia and interactivity to a small extent. These stories are therefore most reminiscent of the view of feature journalism held by the group of journalists who had not previously worked in print newsrooms.

How did *Magasinet online* look in 2009? The section had now changed in appearance. It no longer resembled the print *Magasinet* and did not distinguish itself from the online newspaper's other sections in appearance. But the articles were to a great extent constructed following the same pattern as in 2005. Two characteristics had, however, changed significantly since 2005. First, the articles were not as news oriented. 42 per cent of the articles were either news or adhered to the news flow compared to over 60 per cent in 2005. A lot more of the articles could be categorised as timeless (43 per cent compared to 25 per cent in 2005). Second, the articles contained indications of online discourses to a greater extent: all the articles were furnished with an opportunity for readers to leave a comment. At the same time, about one in four articles was furnished with multimedia content (mostly video) other than photographs.

The content analysis shows that the texts from 2005 and 2009 are fairly similar. If we compare the findings to traditional feature journalism, the most striking feature is that the discourse of intimacy is almost absent. This is somewhat surprising as this discourse was emphasised by the journalists in the interviews. However, as almost none of the articles are reportages where the journalist has engaged in face-to-face interactions with sources, it is understandable that the journalism lacks the kind of intimacy that such encounters would promote. The lack of reportage also weakens the discourse of adventure, simply because it is difficult to provide readers with adventures when the journalist does not leave her desk. This was expected based on the interviews but is still somewhat surprising, as the journalist who ran the section alone in 2005 expressed a wish to have a colleague specifically so that she could write reportage to a greater extent. In 2009, when she had gained two new colleagues, there was still no more reportage. The greatest change, which the increase in resources had brought, was that the rate of publication had increased. The 30 articles from 2005 were published over a period of two months (4 April–2 June), while the articles from 2009 were published in the course of just over a month (1 January–6 February).

However, the literary discourse had a strong presence as a clear majority of the articles were written according to narrative principles. This also corresponds well to the findings from the interviews. Another central feature of the developments between 2005 and 2009 is that online discourses had gained a more central role, with extended use of interactivity, but also multimedia. At the same time, the news discourse had been given less space.

The question is therefore *why* the texts are as they are. What is it that has influenced their form and function, and how they have changed? To begin answering this we have to take a look inside the newsroom.

Social Context

10 May 2005, 9.30 A.M.: 12 journalists, a shift leader and a news editor sit round an oval table next to the hard-copy newspaper's news desk. They form the editorial team of *dagbladet.no* and they are holding their morning meeting. The shift leader, who is chairing the meeting, mentions several things they might work on, but no one shows any enthusiasm.

"It's the Norwegian Mineral Water Championships today,"[3] he says and looks at feature journalist Astrid Meland, editor of *Magasinet online.*

"You can just forget it," Meland replies before saying that she is working on an article on a new film—a film about a man who passed himself off as film director Stanley Kubrick for four years.

"The Japanese royal family is coming to Norway today. That might be something for *Magasinet online,*" another journalist says. Meland nods and says she agrees before adding:

"But I haven't got time. So you'll have to do it yourself."

After the meeting they all go back to their desks. The editorial team sits in an open-plan newsroom in a corner of the hard-copy newspaper's newsroom. They sit in groups according to the section they work on. Astrid Meland sits in the middle of the online newsroom. She runs *Magasinet online* alone. She writes articles, finds photographs, edits, publishes and handles readers' feedback. Every Friday she also agrees what should be published online from the hard-copy edition of *Magasinet* with its editor. She also tries to get the others in the online editorial team to contribute material. They would like to but rarely have time. Meland says she has an ambition of publishing at least one item on the section every day. She writes most of them herself, but about once a week she gets an article from the others in the editorial team. She complains that she sits at her desk far too much.

"I would rather go out, but it is difficult as long as I'm alone. I hope to have one more journalist appointed so that I can go out and write more reportage," she says.

She got the idea for the Stanley Kubrick impersonator story after watching a programme on Swedish televison. The item is finished at 10.30 A.M. She publishes it on the section and has sent a proposal for a front-page teaser to the shift leader. The shift leader publishes the teaser as the lead story on the front-page of *dagbladet. no*. At 10.55 A.M. Meland checks "the log"—an overview in which the journalists can see how many readers are clicking their way to the various stories. She wants to see how the Kubrick article "is going," as she puts it. The log says 6,500 clicks—a number that indicates that an average of 6,500 will click on the story in half an hour if the traffic continues as it is.

"It's going like crap!" she shouts at the shift leader.

She then checks her email and sees that she has received 15 emails from readers about an error in a caption. She corrects the error, but the emails are flooding in. In two minutes she has received 21 emails.

A colleague suggests that the front-page headline could be slightly changed to make the story "go" better. He believes the headline needs a stronger signal phrase, for example "film genius." The headline is now the same as it is on the article itself: "Acted as if he were Kubrick for four years". Meland logs onto the front page and changes the headline to "Acted as if he were a film genius for four years". A quarter of an hour later, the log number has increased to 8,800. But this is not enough to keep it as front-page lead story. At 11.24 A.M.—an hour after it was published—the story is replaced by another lead story on the front page.

Meland doesn't look bothered—she's already working on a new story.

The description above is typical of how the editorial team of *dagbladet.no* in general, and the journalist working on *Magasinet online* in particular, worked in 2005. The online newsroom was small and the online identity was strong. The feature journalist was regarded as a steady supplier of lead stories to the front page, and the pressure to produce stories was great. The focus on the readers was strong, headlines and lead texts were rewritten if the stories were not being well read. The newsroom was strongly characterised by a constant chase for news, there was great competition with other online newspapers and websites. The journalists rarely went out.

In other words, the production of *Magasinet online* was highly influenced by the other online journalists' discursive practice. This meant that the discourse of immediacy was also dominant in the feature journalism, as the content analysis of the texts from 2005 in particular showed. At the same time this resulted in the feature journalists not seeking out sources and milieus outside the newsroom, which marginalised both the discourse of adventure (the opportunity to write reportage) and also partly the discourse of intimacy (the opportunity to meet people and visit milieus at close quarters).

This editorial work culture changed only to a small extent over the years, even though the newsroom gradually grew larger and became more spread out. However, the online feature journalists' role became dependent on managing to distance themselves from both other online journalists and the journalists at the print *Magasinet*, i.e., traditional feature journalism (Steensen, 2009b). They did this by defining their journalism as knowledge based, rich in sources and in depth in contrast to the general perception of online journalism as simple and of poor quality. And they placed importance on news and dialogue with the readers rather than the more reportage-driven and immersive journalism of the print *Magasinet*.

This provides valuable context for several of the findings in the content analysis. First, it may explain the very extensive use of sources and links. It was obviously important to the online feature journalists that their stories should stand out as being in depth and investigative. The number of sources and links became a means to achieving this. But because many of the sources were secondary and many of the links led to the same sources, this can be interpreted as if it were more important to create an *impression* that the journalism was in depth and investigative than it actually being so.

Second, it may explain why the section's appearance was changed from resembling the print *Magasinet* to being the same as the other online sections. Third, personal involvement was interpreted in relation to the readers rather than in relation to the sources as is usual in traditional, more reportage-driven feature journalism. The journalists should involve themselves in the debate that arose in the wake of the articles. They also implemented this, at least to some extent. One of the online feature journalists in particular developed the habit of participating in the discussions that arose in the comments field beneath the articles after she had published one.

This may be interpreted as a continuation of feature journalism's traditional discourse of intimacy. The online feature journalist's expectation of personal participation is not manifested in the actual texts they produce, but afterwards, in contact with the readers.

In 2008 the whole of *Dagbladet* and *dagbladet.no* moved to new premises and the online newsroom became even more spread out. *Magasinet online* was left to its own devices within the online newsroom to a much greater extent and was not part of the constant chase for front-page lead stories and immediacy in reporting as it had been before. This may explain why the news focus in 2009 was no longer as strong as it had been in 2005, as the content analysis showed. However, the work culture was generally the same as before. The feature journalists' everyday work consisted of sitting by their computers, and the practice found in 2005 had to a great extent become routine and therefore institutionalised in the work culture of *Magasinet online*.

Conclusion

A key objective of this chapter has been to show how textual analysis can and should be combined with analysis of other elements of discursive practice in order to understand how genres develop as they do.

The findings show that the journalism of *Magasinet online* is more news driven and reader-involving and less adventure oriented than traditional feature journalism. The analysis of the discursive practice on *dagbladet.no* provides possible explanations for why it became so. The feature journalism on *dagbladet.no* was shaped to a great extent by a need to distance itself from the journalistic practices in which it originated; both the print feature journalism of *Magasinet* and the journalistic practice of the online newsroom as such. The social function of the feature journalism they produced therefore seems to have changed in the direction of providing investigative reporting and engaging the readers in public debate on societal matters.

Does the feature journalism of *Magasinet online* constitute a new genre? To answer that question it would have been necessary to include analysis of how the texts were received by readers in order to pinpoint exactly what kind of social function the texts establish. However, it seems reasonable to conclude that the journalism in *Magasinet online* is the result of what Fairclough (1995) calls a creative discursive practice—i.e., a practice that creates texts in which various discourses and genres fight over hegemony. The discursive practice of the online newsroom clearly affected the production of online feature journalism. The texts are perhaps best characterised as hybrids, as something with both generic and discursive characteristics of both traditional feature journalism and mainstream online journalism. It is therefore difficult to assess what kind of *exigence* they address, and therefore they do not constitute a clear-cut genre.

Such hybrid texts are markers of societal change, according to Fairclough (1995). Further studies of such texts *and* what shapes their production might therefore provide insight into how, and why, society changes.

ENDNOTES

1   The empirical material is systemised and analysed using the qualitative data analysis software HyperResearch 2.8 and Excel. For a more in-depth account of the methodological approach, see Steensen (2010).

2   According to interviews with the editor of *dagbladet.no*, Esten O. Sæther, former editor of the print *Magasinet*, Gunnar Blenness, and the journalist who was responsible for *Magasinet online* during its first years.

3   This quote and the following are translated from Norwegian by the author.

CHAPTER EIGHT

The Process of Covering the 2008 US Presidential Election at Salon.com and LATimes.com

Brooke Van Dam

This study closely follows two news organisations in their reporting of a particular event: a useful approach to news production ethnography rarely encountered in the context of new media. This comprehensive research with the LATimes.com and Salon.com during the 2008 US presidential election illuminates how different organizational structures and priorities impact the resulting online journalism. While the legacy online newsroom was conditioned by its relationship to the print newsroom, the net-native was characterized by the autonomy of its reporters. The push for immediacy in the first case led to chaotic coverage, while the more analytical approach of Salon.com led to the neglect of some topics which had already been reported by other media.

EDITORS' NOTE

The 2008 presidential election was one of the biggest news stories to come along in quite some time due to its historical nature, the prominence of America in world politics and the personalities that ran for the highest office. This study breaks down how two news websites covered this election beginning from how the newsrooms are set up to how the process actually looked while news was happening. The first is the online presence of one of the most recognizable US brands in newspaper journalism, *The Los Angeles Times*. It is a big organization going through financial turmoil as the newspaper market in the United States is losing ground due to multiple economic factors (PEJ, 2009). Salon.com is a net native, independently run website with a small staff and a lot of online clout.

The way in which the two websites go about creating content provides a significant contrast in online newsgathering and approaches to covering a story of national significance with many potential angles and storylines. This study brings together key findings that emerged throughout the five months of data gathering at the LATimes.com and Salon.com. These themes include contrasting physical spaces in creating online news, the networked and autonomous structure within newsrooms, and a need to constantly publish new information which all leads to a very uncontrollable news narrative.

Methodology

The Salon.com newsgathering organisation is divided into three bureaus: San Francisco, New York City and a much smaller operation in Washington, DC. I tried to spend an equal amount of time in the San Francisco and New York offices, three days each, but due to an editor's changing schedule only spent two days in the San Francisco office. While this was a minimal amount of time, it provided a complete picture due to the smaller nature of the operation and the access I was given.

The LATimes.com newsgathering organization is much more centrally located. The main building and newsroom are located in downtown Los Angeles. There are other bureaus and reporters throughout the United States and in twenty other countries, but the great majority of the staff is located in LA. In total I spent eleven days in different parts of the large historic building but all of them within editorial. The final part of my studies included in-depth interviews with over thirty key journalists at both news websites. I also formed a friendly relationship with various people at the sites, which allowed me to gather information over time through informal conversations.

This research is rooted in the cultural chaos theory of media presented by Brian McNair (2003, 2006). McNair sees journalism as influenced by a variety of factors that move us away from the control paradigm (Curran, 2002), which stresses the media's subjugation to authority that has dominated theoretical understandings of journalism for so long. In that, cultural chaos is a direct challenge to the dominant critical theories (Hall et al., 1978; Tuchman, 1978; Herman & Chomsky, 1988) that have underpinned many of the studies of newswork.

McNair sees the current model of media as non-linear with constant feedback and adaptation as new cycles evolve. By using a paradigm based in the natural sciences, he thus seeks to end the grip the control theorists have had over explaining journalistic production. The chaos theory implies an ecological or environmental model of media production. It abandons reliance on a control model that is often used to explain most media production today. McNair states: "more media, moving more information further and faster, means a more chaotic communication environment, with corresponding implications for the acquisition and management of power in society." (2006, p. xx) Although McNair is speaking on a macro-level, I intend to test his theory on a micro-newsmaking level wherein many environmental factors go into making news.

The Two Websites

Both LATimes.com and Salon.com are formidable names in the world of online journalism, yet the background of each and their reasons for publishing online could not be more different. The *Los Angeles Times* brand has thirty-nine Pulitzers, almost a dozen online journalism awards, over one hundred and twenty years in print, the fourth largest circulation in the United States, and claims to be the largest U.S. metropolitan daily newspaper. And yet, the *Los Angeles Times* newspaper also finds itself facing shrinking circulation and is at a distinct loss as to what is next. Additionally, the corporate owners of the *Los Angeles Times* (the Tribune Company) were involved in bankruptcy proceedings beginning in 2008 and continuing through 2010.

David Talbot began Salon.com in 1995 with a few of his colleagues from the *San Francisco Examiner* newspaper. It was one of the first net-native news organizations as the World Wide Web began to change the way we get news and information. It soon won many Internet-related awards and received good press. Since then, the site has survived the dot-com bust (1999–2000), which saw it ask for financial donations from readers in order to survive, and the departure of its founder in 2006.

Physical Spaces and Structure of News Gathering

One of the central tenants of the control theory, used to describe the logics of news production, is the idea of hegemony introduced by Antonio Gramsci (Gitlin, 1980). "Hegemony operates culturally and ideologically through the institutions of civil society which characterises mature liberal-democratic, capitalist societies. These institutions include education, the family, the church, the mass media, popular culture, etc." (Strinati, 1995, pp. 168–169) Gaye Tuchman (1978) claimed that sociologically speaking, there is no conflict between the professional and the organization in newswork. They ultimately both serve to legitimate the status quo.

The structures that were put in place around television news programs or newspaper publishing deadlines may have reinforced an organizational allegiance to a certain way of going about creating a news product (Schlesinger, 1978; Fishman, 1980). However, what was occurring in these two online newsrooms/hubs, bore little resemblance to a controlled environment. On the contrary, in their own way, the two newsmaking structures were very dissimilar to processes and routines that came before.

Networked Chaos: LATimes.com

One of the first things I was made aware of upon entering the Los Angeles Times building was that the "Web Team" was separate from the content being created by the various news desks. This was most obvious in the fact that the Web Team was located two floors above the main newsgathering operation. The Team consisted of a top executive editor, several content and technical managers, an interactive team primarily in charge of building databases, homepage producers, a blog editor and Web deputies. Web deputies were placed on each of the news desks (Foreign, National, Business, etc.) to try and help the newspaper reporters and editors get content to the website quicker and with more efficiency.

Blogging at the website is structured in a few different ways, depending on the subject matter and whether it is singularly authored or done by joint authors. The Top of the Ticket blog, which is authored by two national correspondents, is primarily run by the National desk. This blog was started to cover all of the campaign news for the 2008 election.

This network structure of getting news to the Web relies on technology, a Web department, a specific news desk and finally the visuals department. On a day-to-day basis the result of this structure creates results that are never the same and often represent the chaotic environment that McNair (2006, p. 50) describes.

The tension between daily publishing deadlines of newspapers and the constant need for new information on the Web is one of the largest problems within legacy newsrooms (Boczkowski, 2004a). This news gathering operation has chosen to keep the majority of their reporters working around deadlines created for the newspaper and although stories can be rushed or finished early in order to get them on to the website, that is more the exception than the rule, particularly in the case of the 2008 election. Almost all of the election articles created by Los Angeles Times reporters were put up on the website in the late evening in conjunction with their newspaper deadlines. They did devote one reporter to breaking election news on the website, but in all my observation he seemed to work on updating stories rather than creating new content.

The website chose to use larger syndicated news gathering operations such as Associated Press, or the website's approximately fifty blogs, to deal with the constant need for new information on the web. According to the LATimes.com executive editor: "Blogs are the center of gravity for original content on LATimes.com. We do a lot of updating throughout the day, where we get new stories on the bailout or the status of the debates or whatever it is, but the blogs are really where we have a lot of people." The site also handles breaking news through photo galleries, which are very popular, as well as video or discussion boards.

By the end of my ethnography, the Web Team was moved down to the floor which contained the news gathering desks.[1] Despite this, there was still a general lack of priority and focus given to the website. When it came to the election, it seemed nearly impossible for those in charge of the Web coverage to find extra people to help out with a particular story, and the lack of top editors at any of the Web planning meetings was conspicuous.

The coordination of election content was done by the National desk. They had many reporters covering the campaign from numerous angles. The desk had a reporter travelling with both Barack Obama and John McCain almost all of the time. In addition, they had a full team of at least a dozen reporters in Washington DC as well as a smaller presence in the Los Angeles bureau. Most of the coverage provided by the National news desk was planned around the newspaper, and then the website put up whatever was published. It was very strategic and involved a lot of voices from interns all the way up to the top editor of the newspaper. In this sense, it could be argued that what the *Los Angeles Times* was doing for the website mainly amounted to shovelware (Pryor, 2002). However, the Web Team did provide additional content that made a mixed bag of shovelware, original material (in the form of its blogs, video segments and photo galleries), and aggregated items from other syndicated services.

The Top of the Ticket blog was a key platform for the election. It was featured heavily on the homepage and became a presence in its own right in the blogosphere.[2] The blog was kept up to date by two reporters, one based in Los Angeles and one in Washington, DC, who were given almost total autonomy in what they wrote. Nothing had to go through content editors but each post was copyedited. In addition, a National desk researcher added posts. Other reporters would put up posts, but that was usually to tease a larger article they were working on for the newspaper.

The photo galleries were given huge priority on the homepage and were used many times to tell a story, such as the debates or 'A Day at the Conventions.' Despite having multimedia video-producing staff, video was rarely used as a centerpiece to a story but rather as an add-on to other more prominently featured content. The Web team linked to live streamed video during the conventions for prominent speeches. The interactive team produced electoral maps, speech tag cloud bubbles and voting registration platforms for users of the site to engage in the process;

however except for on election day these were not featured as prominently on the homepage or separate section created within the website called Campaign '08. The way the newsroom at the *Los Angeles Times* went about covering this story was quite similar to other news. It was a highly fragmented network due to the stratification of managers and no central leader. The executive editor of the website had complete control over LATimes.com but had very little say in what each news desk was doing. The biggest difference for the election story was the amount of resources the national desk gave to newsgathering on the topic as well as the emphasis on the Top of the Ticket blog. The blog became central to the online coverage as it provided a reason for users to come back to the website on a regular basis.

Autonomy for Journalists: Salon.com

Salon.com is not a 24/7 news operation, viewing itself as a Web magazine. It deals with original stories in two principal ways, either in traditional text-based article form or through their primarily text-based blog content. In addition, after their user-generated content blog site Open Salon debuted in September, occasional articles were put on Salon.com. The commissioned articles are published once a day in the evening, which is how they have chosen as a daily Web magazine to deal with production deadlines. One of the articles is chosen as their 'cover' story for the next day and given prominence on the homepage. They have a number of blogs on their website that deal with anything from feminist issues (Broadsheet) to economics (How the World Works) as well as comedic cartoons and video content. The War Room blog was central to election content as it was their fully devoted politics blog.

The presidential race had a huge presence on Salon.com both because it was a part of the Salon brand (according to its editors) and because it contributed to an increase in the site's popularity. Several of the editorial staff mentioned to me that the "bread and butter" of the site was its political coverage. According to a project manager: "I think most of our traffic tends to go to political stories."

The bulk of the political coverage was handled through the news editor. He was primarily in charge of the three correspondents who were based in Washington, DC but found themselves travelling across the country much of the time. The news editor also worked with the key War Room blogger who was based in the New York office and any additional freelancers who wrote for the blog. He would also coordinate with weekly or monthly columnists who wrote pieces for the site and would edit and commission stories by freelance writers.

The economics blog How the World Works also became heavily infused with politics in the latter part of the campaign. As the economic crisis hit in September, the blog increasingly wrote about how the candidates were responding to the downturn and analyzed their solutions to the problem. The blog, which is always available on the website, was given higher prominence on the homepage and often given 'cover' status as well.

Salon.com offices, both in San Francisco and New York, are located in large office buildings in the center of the respective cities. The desks are only half occupied. Most of the reporters, and even the managing editors, are not required to be in the office or have a set schedule. It is a very autonomous work environment. However, with this autonomy comes little camaraderie and sense of a news team. Obviously, these people work together to create an end product, but there is little in the way of verbal discussion and the newsrooms are very silent for large portions of the day.

There are various meetings throughout the week that are all done via conference calls[3] due to the dispersed physical location of staff. The subject matter of election content was primarily handled during the Friday afternoon "news meeting." The meetings were short and moved quickly from topic to topic. The managing editor for news asked what the different correspondents were working on and made some suggestions for future stories. They also discussed upcoming articles that have been commissioned by freelancers to avoid overlap in content.

It is not a rigid process in deciding what gets covered and what does not. The site will make sure to cover what they feel are the big stories in politics and culture, but beyond that the reporters have a lot of autonomy in what they want to write about and in pursuing their own leads. According to one of the political reporters: "You know occasionally we'll get an assignment from an editor but that's pretty rare. Usually I could do mostly whatever I want, which is kinda nice." He is in constant communication with his direct section editor but that relationship is casual and fluid. The various editors will periodically fly to the different offices in order to interact with the other staff members.

The process and structure in this environment gave the journalists and bloggers a high amount of autonomy. The use of individually authored blogs, which were given much more prominence and used more frequently to tell the election story than on LATimes.com, prevented any coherent editorial line and allowed specific voices to find a niche in the online environment. This set-up is exactly what McNair describes when assessing chaos in light of news organizations. "It views journalistic organizations and the professionals who staff them as more independent and disruptive of power in their communicative activity than their allotted role in critical media sociology has allowed" (2006, p. 4).

As News Happens

This section seeks to highlight how the structures and routines that were put in place work themselves out as news happens. As detailed above, LATimes.com and

Salon.com have chosen very different ways to go about structuring their newsrooms and getting content to the website. What this creates when news is happening highlights both the problems and the strengths of both websites.

The Nomination of John McCain: LATimes.com

The final day of a party convention is the moment when the nominees formally accept their nominations for president by their respective parties. An event like this, with planned coverage by the journalists and desired control of the narrative by the politicians, provides a backdrop to see how a live unfolding story plays out in a newsroom. The nomination of John McCain in particular highlighted the competing factors that lead to an uneven and highly chaotic atmosphere of news production at LATimes.com.

At around 2 P.M. Pacific Time on the day McCain was accepting the nomination, and the final day of the Convention, the Web deputy still did not have a schedule from the Republican Party for that night's speakers. She told me this had been the case for most of the convention and that the reporters were forced to have conference calls with Republican officials each morning in order to find out the schedule of events. The Party would then send email updates throughout the day, changing the schedule.

There were many technological issues that played into the election coverage on the day of McCain's nomination. The Web deputy informed me that after purchasing Flip Cams[4] for the reporters and bloggers, most of them were not being used. There was a constant sense of frustration by much of the Web Team editors about the perceived disinterest in other platforms by the newsgatherers on the National desk. Many of the reporters for the National desk had a hard time figuring out the content management system and had no interest in posting on the blog. This technological divide only reinforced a lack of convergence within the newsroom and created an online versus newspaper situation.

This lack of enthusiasm for multiple-platform journalism also created workload problems. I watched the Web deputy handle a majority of the website coverage during the conventions because the reporters and editors at the National desk were not concerned with the website. They were instead, consumed by newspaper deadlines and therefore chose not to contribute to online coverage. The Web deputy remained in the office until 4 A.M. many nights in order to update coverage on the website *and* to "fix the kinks," as she would say.

The team of homepage producers had a clear focus on getting news to the website as soon as it was humanly possible (Domingo, 2008a). As each convention day progressed, the frustration with the National desk became stronger. One of the producers confided to me that they had a problem most of the week with the speed of stories being delivered. They were not getting stories or blog posts quickly enough to put up on the site and in order to keep up often used photos from a photo agency or an article from the Associated Press.

There was also a sense of frustration within the Web team regarding how many people a story had to go through when it came from one of the desks downstairs. One homepage producer noted that the amount of people that have to "touch a one sentence story is crazy." It became such a problem that many of the Web staff as well as the executive editor for LATimes.com decided to have an *impromptu* meeting to discuss the problem. They talked about how to compete with other seemingly faster news websites such as NYTimes.com, but had no power to make reporters publish quicker there was no ultimate conclusion.

The Web team tried to solve this problem on its own on the night of McCain's nomination. Senator Lindsey Graham, former Governor Tom Ridge, and Cindy McCain (McCain's wife) would all be introducing the Republican candidate, and the National desk had already told the Web producers that they would have no content for them during those successive speeches. It was decided, less than an hour before the event, by the executive editor of the website, the Web deputy and two homepage producers that they would attempt to live blog the three back-to-back speeches.

The Web deputy instantly began trying to recruit a National desk reporter to live blog the speeches. Everyone on the National desk told her they were too busy and to "have the Web people do it." The two main bloggers who wrote for the site could not live blog, according to the Web deputy, as they were working on other blog posts. It was eventually decided the duties would be split between the Web deputy and one of the homepage producers.

Various people then began coming up to the Web deputy, and in the midst of several conversations she looked up at the television screen on her desk and realized that Senator Lindsey Graham was taking the stage. Amidst some confusion, she and the homepage producer upstairs began to send instant messages to each other to try and figure out what to do. The homepage producer quickly began blogging. The Web deputy then went into the post and found a live video link to attach and was trying to add speech text in order to get it published and up on the blog. As she was trying to do this, the Top of the Ticket blogger began calling her cell phone numerous times due to problems he was having.

After that was sorted, the Web deputy went back to the live blog, but by now Senator Lindsey Graham was off the stage and former former Governor Tom Ridge was about to walk out. She quickly published the Graham live blog even though it was no longer live and set up the next live blog for Ridge's speech. The deputy spent a lot of time trying to sort out what to say and finally published the post mid-way through his speech. After this was completed, she went back into the Graham post to take out the live video link and fix some grammatical errors. She then told me that she felt this maybe was not the most efficient way to go about things. It was too much information too quickly for one person to write about in any meaningful way. So instead of updating the Ridge blog, she began straight away on the Cindy McCain speech, constantly coordinating via instant messenger with the homepage producer upstairs.

Once Cindy McCain was done speaking, the Web deputy called upstairs to one of the homepage producers in order to make sure they were on the same page as John McCain was about to go onstage. The National desk had produced an updated story that was ready for the website as soon as Senator McCain walked out. All of this content was the centerpiece of the homepage during the period it was occurring.

The confusion around how to handle even pre-planned events like the debates was the consequence of the networked structure LATimes.com created. Although all of the teams were coordinating with each other, with no central leadership and no priority given to the Web, individual players were left making last-minute decisions based on technical or personnel factors.

The John Edwards Affair: Salon.com

The atmosphere in the news hubs of Salon.com contains much less verbal and face-to-face discussion and confusion. Watching an election story unfold is almost impossible to capture because of the large majority of communication happening over instant messenger and email.

There was very little in terms of pre-planned election coverage that I was able to observe. However, during the New York observation period former presidential candidate John Edwards came to the forefront of the narrative again when he admitted to having an affair. The process of covering that affair showed the quick nature of Salon.com's decision making as well as their slower and more traditional approach to writing articles.

The newsroom was eerily quiet, as usual, when the New York editorial director came out of his office and into the open area of desks and announced loudly that: "John Edwards admitted he cheated on his wife. ABC News just posted it on their website." The small news staff began casually chatting about the story that had originally been given no serious coverage by most of the mainstream press, even though the tabloid newspaper *National Enquirer* had been reporting it for months.

The War Room blogger posted the information on the blog soon after a brief discussion, linking to the ABC News story. As soon as the post was up, it was highlighted on the homepage by another editor. The news editor then called the Washington, DC Bureau chief to inform him of the breaking news and told him to follow up on the story.

Despite the story being uncovered on a Friday, the news editor decided that further article coverage and angles on the story would be discussed the following week. The news team conducted a conference call a full five days later. The news editor and two feature editors in New York, even though they only sat a few feet from each other, dialed into the telephone conference system to chat with the editors at the other newsrooms. They spent a long time discussing various story angles Salon.com might take but ultimately none of these ideas became articles as most of the team felt the story had been covered extensively on the website and by other news outlets already. They felt it was not worth their time and resources to add further to the story.

Many of the potential stories they were discussing had already been addressed in posts on the War Room blog. The only two longer form articles Salon did went up on the website over a week after the news broke.

This incident was indicative of the approach Salon.com had to doing its journalism. With such a small team and the level of autonomy given to bloggers, they were able to post elements of a story quite quickly in blog form but took longer to write articles. The underpinnings of Salon.com, which was set up by former journalists for the purpose of creating a place for discussion, bled into much of the autonomy it gave its reporters and bloggers. There was no mistaking that although the editors were in charge to a degree, the reporters were not being told what to do.

It also shows the importance of other media as a source for information and the networked structure of information sharing that exists in journalism. McNair compares the flow of information on the Web with storms. He says:

> ...the network structure of the World Wide Web, in combination with the 24-hour presence of real-time satellite news, produces an environment where information cascades become more unpredictable, more frequent, and more difficult for elites to contain when they begin. News storms develop without warning, placing power elites on permanently reactive, defensive mode (2006, p. 202).

Conclusion

The legacy news website LATimes.com and the net native news website Salon.com provide an excellent case study in contrasting ways to go about covering the same story. With the backing of a large newspaper, LATimes.com provided a lot of content for the user but very little in the way of a coherent structure as to how to cover a story online. It was rather the case of haphazardly producing as much content on as many platforms as possible and trying to highlight it to the many users of the website.

In contrast, Salon.com offered much less content and relied heavily on text-based blogs to tell their story. Their newsmaking structure was autonomous and relied little on verbal conversation between a lot of players and a lot on digital communication between a select few. What we are seeing with these two websites is close to what McNair describes as cultural chaos. "The chaos model stresses unpredictability of outcome in media production processes, a consequent uncertainty around the quantity and quality of information flow, the importance of feedback loops, and enhanced volatility in the management of both communication and power" (2006, p. 49).

ENDNOTES

1   The interactive team, which was a part of the larger Web Team umbrella, was moved to the second floor and situated next to many of the design desks as well as the videographers

2   The blog increased its page views from around one million in June 2008 to over four million in October 2008. It also consistently appeared in the Technorati Top 100 blogs around the time of the November election.

3   Each journalist calls in to a special phone number where they are then able to talk to all the persons involved in the meeting on a secure network.

4   A brand of small video cameras that fit in the palm of one's hand.

PART THREE

Beyond the Newsroom

CHAPTER NINE

Studying User-generated Content at the BBC
A Multi-site Ethnography

Andy Williams, Karin Wahl Jorgensen and Claire Wardle

The British Public Broadcasting Organization (the BBC) is by many accounts the world's largest news organisation and is certainly among the most influential. This chapter details the efforts of a research team to address the complexity of the news organization—spread across different locations within the UK—when trying to assess the management of audience participation, one of the most fascinating developments in online journalism. The chapter documents the ways in which different news teams of the BBC have adapted to, integrated, or occasionally resisted user-generated content as a new element in the news production process.

EDITORS' NOTE

Over the past decade, technological change and shifting priorities among news organisations have enabled new modes and genres of audience participation, often discussed under the umbrella of "citizen journalism" (Deuze, 2008b, p. 107). What unites most scholarly accounts of these phenomena is an emphasis on describing an increased role for the public in producing material which would formerly have been the preserve of professional journalists. Another common scholarly theme is the idea that the communication of news is no longer a one-way practice and is instead becoming more collaborative and consensual: less like a lecture, and more like a conversation (Kunelius, 2001, Gillmor, 2004). Put simply, the new journalism "must be seen as a praxis that is not exclusively tied to salaried work or professional institutions anymore" (Deuze *et al.*, 2007, p. 323).

However, ethnographic research into the institutions of news production has consistently found that the norms and routines which underpin journalism are influential and difficult to change (e.g., Paterson and Domingo, 2008; Boczkowski, 2004a). Pioneering newsroom ethnographies in the 1970s showed how important the institutional and professional routines of journalism practice are in shaping news content (Gans, 1979; Schlesinger, 1978; Tuchman, 1973, 1978). As Gaye Tuchman (1978, p. 12) argued, "newswork transforms occurrences into news events." The routines associated with this work are durable at least in part because they are central to the continuing authority of journalism: they not only construct external events as news but also constitute self-legitimating practices. Such research has also shown that news production has often been characterised by indifference, and sometimes hostility, to the audience. For example, Gans (1979) established that journalists are largely dismissive of feedback from audiences because they see it as unrepresentative, often coming from a conservative segment of the citizenry, and sometimes bordering on the lunatic (pp. 230–235). This well-documented resistance to the audience is understandable given the fact that journalists' professional status and autonomy often rests on their perceived ability to make more valid decisions than consumers of the news (Schlesinger, 1978, p. 111). It nevertheless clashes with a long-standing emphasis on participatory media formats as playing a key democratic role tied to notions of public service. As such, the desire for professional independence has always co-existed in tension with aspects of journalistic self-understanding which emphasise public service (e.g. Carey, 1987).

Unsurprisingly, given the potential threat to ingrained professional values and practices posed by the rise of audience participation, research has found that mainstream news organisations have tended to approach the challenges of participatory journalism conservatively, often incorporating them within existing routines and practices (Deuze, 2003). Jane Singer has written persuasively about the ways in which the mainstream media have set out to "normalise" participatory media formats until they become subsumed within "traditional journalistic norms and practices" (2005, p. 173).

Scholars have pointed out that the mainstream news media have been "slow to respond" to citizen journalism (Thurman, 2008, p. 139). Alfred Hermida, writing about blogging at the BBC, found that journalists still see blogs as "an extension of, rather than a departure from, traditional journalistic norms and practices" (2009, p. 2). Similarly, Singer (2010) suggested that UK local and regional reporters believe that unless it is carefully monitored using audience material "can undermine journalistic norms and values," and Harrison (2010, p. 250) discussed BBC journalists' continued perceptions of a "very real and quite genuine worry about the threat UGC poses to editorial values and ultimately to news standards." Others have found that opportunities for mainstream media audiences to participate in the processes of news production are often circumscribed (Domingo et al., 2008, pp. 337–339), or "rejected as a form of 'real' journalism" by reporters (O'Sullivan and Heinonen, 2008, pp. 367–368).

This chapter, based on a multi-sited ethnography at the BBC, picks up on these questions to examine how user-generated content is viewed across the organisation by a range of different kinds of newsworkers. In the main, our research confirms much of what other scholars have found: It indicates that most news journalists and editors see audience material as just another news source whose processing is embedded within the long-standing routines of traditional journalism practice, a formulation which is perpetuated by the institutional frameworks set up to elicit and process audience material as well as the content of the corporation's UGC training (Williams, Wardle, and Wahl-Jorgensen, 2010).

Nevertheless, our chapter also reflects on how our engagement with several different sites and locations within this complex news organisation revealed a more complicated picture: We found a broad division between those journalists who primarily produced "hard news", and some of those on the margins of BBC News' operations. Radio, television, and online news journalists tended to value audience material principally for its contribution to news-gathering (in the form of directly submitted eyewitness audio-visual material or tip-offs from the public). Other journalists, for instance those working on the more features-driven BBC Local websites or new media professionals with a focus on building stronger links between the BBC and the communities it serves, were enthusiastic about the potential of new forms of engagement to enable audience creativity and to allow the BBC to engage audiences in extended conversations and collaborative journalism projects. It is this fault-line in journalistic approaches to audience material that we will examine here. This division was not uniform or simple. The picture was complicated by factors such geographical region, diverse newsroom cultures, as well as individuals' roles, responsibilities and locations within professional and organisational hierarchies (including the platform for which journalists produce content, managerial status and seniority, and journalistic specialism). Some of these factors are cultural and some individual, but all point to the need for ethnographies of online journalism to offer a differentiated analysis of complex news organisations such as the BBC and the diverse discourses, values, and practices circulating within them, which in turn inform varied constructions of journalism, its audience(s) and levels of public participation in news production processes. We next reflect on some of the methodological challenges of providing such a differentiated analysis and how we sought to overcome them.

Method

Newsroom ethnographies offer researchers an opportunity to investigate, through direct observation, the lived experience of journalists on the basis of the "patterned, routinized, repeated practices and forms that media workers use to do their jobs" (Shoemaker and Reese, 1991, p. 85; see also Bird, 2010: p. 10). Ethnographic field-work has also been a "sharp tool for discerning not just the unifying features but the divisions, boundaries, and conflicts of the society being studied" (Born, 2005: p. 15). In particular, multi-sited ethnographies are increasingly seen a way of shed-ding light on the diversity of newswork (Cottle, 2007), and this was central to our research design.

Our ethnographic research was designed on the basis of awareness of the com-plexity caused by the sheer size of the BBC as the largest broadcaster in the world. To begin to describe some of this complexity, we conducted multi-sited newsroom observations with a team of five Cardiff University academics spending a total of 38 days in different newsrooms across the BBC, both at the network, regional, and local level.[1] Access to newsrooms was facilitated by the fact the BBC co-funded our work along with the UK's Arts and Humanities Research Council (AHRC). Researchers spent time at the following locations: BBC Devon Plymouth (10 days); BBC Wales Cardiff (5 days); BBC Sheffield (10 days); BBC Leeds (3 days); the UGC Hub (6 days); *BBC Breakfast* (1 day); *News 24* (1 day); World Service *Newshour* and *World Have Your Say* (1 day); and the BBC College of Journalism "Have They Got News for Us?" course at BBC Wales, Cardiff (1 day).

We had a limited period in which to complete our observations because of the demands of finishing an ambitious multi-method project within a year.[2] For this reason a relatively large research team (of five academics) was deployed to BBC newsrooms simultaneously. The observations took place during the first two weeks of September 2007, and the team members were in frequent contact, exchanging notes and discussing emerging findings. Our research questions were intended to broadly map the views and practices around the use of audience material in the news organisation, but for the purposes of this chapter, we focus on examining what the research showed us about (the different kinds of) BBC journalists' attitudes to the (different types of) material, and how this affected their use of it.

Before beginning our observations we devised a plan on the basis of our central research questions. It emphasised exploratory and semi-structured interviews with journalists, ethnographic mapping of flows of audience material, and observation of day-to-day routines including meetings and exchanges between staff and scheduled shadowing exercises. The team consisted of people with varying levels of experience of newsroom observations: some had extensive knowledge and had carried out many such studies before, and others were only familiar with this method from the relevant literature. In order to overcome these challenges and ensure the comparability of our data, a bespoke ethnography briefing pack was prepared to guide researchers in the field. The pack included a suggested timetable of research tasks to be completed, copies of consent forms, an info sheet that explained to media workers what we were doing in their newsroom, detailed scripts and prompts for exploratory and semi-structured interviews, a list of our research questions to assess our progress, copies of a standard field note sheet, and fact sheets which re-iterate and explain important elements of newsroom observation (e.g. "entering the field," "building rapport with gatekeepers and key participants," "observation and what to observe," "recording observations and taking field notes," "carrying out exploratory and semi-structured interviews," and "key ethical issues in newsroom observation"). Frequent research team meetings and phone conversations allowed us to share concerns and experiences, discuss similarities and differences between our sites and ethnographic patterns emerging from them and thus served as a dynamic way of monitoring the consistency of our methodological approaches.

Our design allowed us to track where and how audience material circulated both within and among newsrooms at regional, national, and network levels at the BBC. This was particularly important because of the need to observe the effective-ness of the centralised system for the management of audience material based around the "UGC Hub" in London's Television Centre but also because of our commitment to capturing data from as wide a cross-section of BBC journalists as possible.[3] While many newsroom ethnographies of this type have focused on one particular privileged site of production, our research attempted to reflect the diversity of practices and approaches to user-generated content across different areas of this complex media organisation. As part of the observations, we completed interviews with 115 BBC journalists with varying degrees of experience and seniority. These interviews form the backbone of our data, but they were supplemented by many pages of field notes and a number of maps which allowed us to trace the progress of examples of different kinds of audience content into, within, and among newsrooms.

Findings

We have written elsewhere about the dominant trends we identified in BBC news-rooms across the UK (Williams, Wardle, and Wahl-Jorgensen 2010). Principal among these was the way in which the discourse of most individual news journalists, their daily professional practice, as well as the BBC's institutional approach to dealing with audience material, have not on the whole disrupted traditional relationships between journalists and audience members. On the contrary, many of the new forms of journalism enabled by the rise of citizen participation have been subsumed within traditional journalistic practices such as news gathering, and the potential for new collaborative partnerships between producers and consumers of news has been recuperated by the BBC and is now mainly understood in terms of traditional and long-standing journalist-source relations. But this is not the whole story, and to suggest it was would risk ignoring the fissures apparent in this dominant narrative. A recent paper by Quandt and Heinonen (2009) shows that there are often differences in approach to audience material between journalists working within single newsrooms. They identify two broad attitudes that reporters hold about audience material: a segregationist approach which favours leaving journalism to journalists, and an integrationist approach in which the audience are regarded as an ally in a journalistic process which is co-creative. Our own data confirm this. Here, we chart some of the fault lines along which these approaches diverge based on close attention to the data generated by researchers at BBC Radio Sheffield and BBC Wales.

BBC Radio Sheffield: The Local Newsroom

BBC Radio Sheffield offers a local radio service to the city of Sheffield and the surrounding towns in South Yorkshire and the North Midlands. On an average day there were 12 news journalists in the newsroom working across a variety of platforms at any one time (one news editor and one assistant news editor, a senior broadcast journalist for news, six news broadcast journalists, two television journalists producing local content for the area's regional news programme *Look North*, and two BBC Local Web writers). In addition, the large open-plan office at BBC Sheffield is also home to other broadcasters and producers working on the station's non-news programming, as well as a small "Audience Team" of journalists who visit hard-to-reach communities in order to promote the radio station and (to a lesser extent) gather and create news and non-news content. The regional Web news team is based at the BBC's hub in Leeds, and a small team of three journalists produces all of the content for BBC's South Yorkshire online news site.

Overwhelmingly the news journalists we spoke to here favour the segregationist approach to audience content (Quandt and Heinonen, 2009). A surprisingly large number had indistinct ideas about what user-generated content actually was, and some had not heard of the term before. When they did talk of audience material, they framed it primarily as a source of news to be processed in the same way as that from other sources. Any participatory or democratising possibilities of working with the audience were an afterthought, if they were mentioned at all. As one senior journalist told us when asked if an increase in the use of audience material had changed his role, "we don't think of it as something different to what we've always done." He also suggested that the presence of the researcher in the newsroom had led him to classify elements of his day-to-day journalism practice as user-generated content *post hoc*.

> You're around this week, and I'm conscious of it. I'm thinking that's UGC, and that isn't. But [...] stories have always come in different ways, you know. They might come in on the email, they might get phoned in, they might come from an idea at a morning meeting, and then [...] we'll look at them and think, right, "Is that worth doing?" News values apply. It doesn't really change what you do.

Audience material was seen by most news journalists as source material to be processed rather than as an opportunity for the public to retain creative control over their output or a chance for journalists to collaborate in jointly producing content. In some cases, even the journalistic use-value of audience material was questioned.

We were repeatedly told that "UGC" is often "more trouble than it's worth" because of the time it takes to "bottom out" material from the audience (i.e., to check it for accuracy), but that journalists increasingly have to follow up on leads from the public because of managerial pressure. On the second day of our observation in Sheffield we shadowed a senior broadcast journalist while he gathered stories for news bulletins. Early on we asked what UGC meant to him. He replied, "to be honest, it's the stuff that comes in from the public that you just know is gonna take ages to look into and get to the bottom of." Later that day he fielded a phone call from a member of the public concerned about a local council's responsibility for recurrent floods in her area, and the journalist suggested this was an example of user-generated story that might interest us. We asked him in what way this example of UGC differed from traditional newsgathering, and he replied:

> Well, it's not really [different] isn't it? It's always been there, really, that kind of thing. But [...] we're encouraged to do this kind of thing more, nowadays. Maybe before we would've got something like this that's a bit complicated from the public and we'd perhaps have ignored it. Now we think, "Ooh, that's UGC, how can we use that?"

Those news journalists who were less sceptical of audience material usually framed it in terms of its instrumental value, often with reference to the idea that it allows access to audio-visual content and story tip-offs which would have been otherwise unavailable. Two recent local news stories were often cited in this regard: a nightclub fire in Sheffield and the Yorkshire floods in the summer of 2007, both of which had resulted in unprecedented amounts of photographs, videos and eyewitness statements being volunteered by audience members.

The few reporters who were enthusiastic about the potential of audience material were located on the margins of the "hard news" operation, features writers who worked at a distance from the demands of meeting half-hourly or hourly deadlines.

Some journalists on the Audience Team had worked in collaboration with members of the public to produce audio diaries on a variety of topics relating to their lives, or used a community bus to visit "hard to reach" communities to work on collaborative projects. One of the Internet journalists had recently taken a sabbatical in order to train disadvantaged school pupils in how to shoot and edit their own news packages on mobile phones, a collaboration which resulted in selected content being broadcast on the BBC Local website. This journalist also managed a small team of community reporters with whom she worked to produce hyper-local news content on a range of different subjects. An example of such coverage is a series of five weekly unpaid fan-authored football diaries about the different teams in Radio Sheffield's catchment area. We observed journalists working on the material and corresponding with these diarists, and were struck by the time and effort taken when explaining the editing processes and asking for changes to these texts.

Football diaries represent one example of the many truly collaborative small-scale features-led citizen journalism projects on BBC Local websites. In one session a journalist complained mildly about how long it takes to make such material publishable but also stated his commitment to instructing citizen reporters in good journalism practice. The work we observed involved editing a draft diary entry which was much too long and had made potentially defamatory statements about a football club manager. Once the process was complete the journalist sent an email back to the contributor with detailed feedback. He encouraged him on the positive aspects of his work and provided clear guidance about how to improve future pieces. When asked about the participatory and educational aspects of this way of dealing with such material the reporter displayed a commitment to the ongoing improvement of these amateurs:

> What's in it for us? Well it'll hopefully make my life easier next time he writes something. [laughter] In a sense though, I guess, there's the development of talent coming through in a few years. These people might be coming to work here with us. But a lot of the benefits are much wider than that…they're educational benefits that are part of our public service remit.

BBC Wales: The National Newsroom

At BBC Wales, which is the headquarters for the broadcast corporation's activities throughout Wales—including television, radio and online provision—the vast majority of audience-related activity takes place in the New Media department. This is located in a building across the road from the main site of BBC Wales, and operates separately from the online news team based in the main newsroom. Aside from New Media, the building houses mostly administrative departments, including the BBC Finance department, subtitling offices, IT training, and a nursery for children of staff. The number of employees in the department is somewhat variable, as many work on short-term contracts or on a free-lance basis, but there were 40 computer terminals in the open-plan office, around half of which were occupied at any one time.

The New Media department was responsible for designing the web-based and interactive content for all news and non-news BBC Wales television programmes. Their plan for 2007–2008, for example, involved work on 30 different websites with associated opportunities for audiences to submit content. In addition, the department also managed the BBC Local sites, housed in five different local offices across Wales. Further, the department hosted the digital storytelling team—a group devoted to innovative collaborative journalism activities, including a community bus like the one we encountered at BBC Radio Sheffield, and several projects designed to allow members of communities the opportunity to tell their own stories through the medium of digital photographs and films.

The collaborative and empowering ethos of the digital-storytelling form pervaded the whole department. Most journalists, like their colleagues in Sheffield, found the concept of "user-generated content" unhelpful but for very different reasons. To the BBC Wales New Media employees, the term unhelpfully implied the journalistic appropriation of user materials whereas they preferred a more mutual understanding of audience's participation in media production and used terms like "participatory" or "collaborative" journalism far more readily than "UGC." Though many had trained and worked as journalists, or had experience in news production or management, they spoke of their own work using terms like "creativity", "innovation" or "community outreach", and juxtaposed their work quite deliberately to that of news journalists. The physical separation between New Media and the rest of BBC Wales was also echoed in the distinctive professional culture of the department. The producer for BBC Local mid-Wales, who used to be a "hard news" online journalist, commented that the BBC Wales main online news operation was like a "sausage factory" and "a bit like Stasi headquarters, all people squeezed into a small dark room, working on their deadlines at 12, 3, and 6, compared to the chilled atmosphere over in New Media, where there are potted plants and lots of space."

Central to the ethos of the New Media Department was an emphasis on the "authentic voice" that can be derived from allowing people to speak "in their own words." As the deputy editor of New Media put it, you "have to let the people tell the story for themselves, you have to leave some gaps for them to fill." He was strongly critical of journalists' "parachuting" into communities to gather ordinary people's stories and then promptly leaving. He and many other interviewees underlined the considerable expense of producing collaborative content: "We've looked at the value for the money sort of thing [and] it's no cheaper than making a programme with a proper journalist, but you get all these other values out of it…well we say, this is a different approach to the approach traditional journalists will take." He described digital storytelling as the "Rolls Royce" of methods, insofar as it enabled the collection of "extraordinary stories in normal life." Similarly, a digital storytelling producer described the form as a "a partnership rather than bossing." He recalled a digital story by a woman who had lost a son from leukaemia:

> The hurt was really fresh and she wanted a way to tell her story and she came and she knew that was her story, it was the biggest thing she wanted....What usually happens is [before] National Leukaemia Week, the Newsroom Editor says, "Get me a story about Leukaemia" you know the shots it's going to be, it's going to be the mother and father sitting on a sofa leafing through the photo album looking sad, and they will point the camera at the mother and ask her enough questions until eventually she'll cry....But Gaynor Clifford who made this story about Leukaemia...came at it from a different angle and she looked at the way that her son, when he knew he was going to die, faced up to his death and prepared for it. And that was really something that we all learnt about.

Others fondly recounted unusual and personal stories collected from audience members, ranging from one about a collection of porcelain owls, a man's recollection of his childhood rock pool lobster fishing trips, and a woman's story of winning a crossword contest.

There were, of course, pragmatic reasons why audience material was valued at BBC Wales. It allowed the organisation to reach distant communities in a geographically complex and sparsely populated country, to draw on local knowledge and report on stories that wouldn't otherwise be told. These strategies contributed to creating a sense that Wales could be seen as a coherent nation, constructed and represented by the BBC. On the BBC Local sites, audience material of various kinds provided the main form of content, and to the producer for South-East Wales it was the "best way you can actually get rigged up to grass roots":

> People like the fact that it's not all written by journalists. They like the fact that it's ordinary people having their say and that's pretty much the main strength of the appeal. But also it's relevant to local, because it's local people who are writing it, it's not, you know, the word from the BBC on high.

The New Media employees suggested that they enjoyed a great deal more creative freedom than their journalistic colleagues across the road. However, even among these, there was a considerable degree of interest in some of the opportunities presented by audience participation. Among other things, BBC Radio Wales had started a "Community Reporters" programme where individuals in different communities across the country were contributing local "soft news" stories.

In general, journalists who worked in the conventional newsrooms of BBC Wales conformed far more readily to a journalism-centric view of audience material. For example, a young television journalist commented on what he saw as resistance among journalists to the use of audience material from an elitist perspective—he argued that journalists are trained to provide quality news and that the use of such content undermines this quality. He also questioned the use of audience material as "a lazy way of newsgathering because it provides you with the voices and content of the people who yell the loudest, but not necessarily the ones who have the best information."

For the "hard news" journalists working at BBC Wales News Online, the main value of audience content was the provision of story ideas and sources, similar to the position of BBC Radio Sheffield reporters. Getting "good examples" of personal stories was also seen as a "great tool" for staying in touch with their community. As the senior reporter for online news put it, "it's good to have people with actual experiences and opinions that are real getting in touch with you and adding to your output." Nevertheless, on occasion, the discourse of collaborative journalism also made its way into the vocabulary of those in the main newsroom, as when BBC Wales' acting editor of News Online, described a "partnership" between journalists and the public.

In many respects, BBC Wales was characterised by a strongly normative understanding of audience material in terms of engaging communities and enabling authentic story-telling through the investment of resources in "collaborative" or "participatory" journalism. This view suffused the culture of the New Media Department but also had an impact on other aspects of the news gathering operation. The journalism-centric positions that dominated accounts from journalists in Sheffield were clearly in evidence—but in Wales, they were more marginal rather than central to newsworkers' practice and self-understandings.

Conclusion

Our ethnographic work took place in very different newsroom cultures which clearly shaped the dominant views and discourses on and around audience material. These were, in turn, affected by the place of the newsroom within the structure of the BBC—where journalists at BBC Radio Sheffield were driven largely by the logic of news-gathering in a relatively deadline-oriented environment, BBC Wales (influenced by the strong collaborative ethic of the New Media department) placed significant emphasis on audience empowerment and collaborative journalism as a way to connect geographically distant, difficult-to-cover communities and provide a sense of Wales as a nation.

As a team of researchers working in these varying environments, it was challenging to look beyond the newsroom in which each of us had worked and see the "bigger picture," revealing dominant trends and attitudes but also teasing out the ideological and practical fault-lines along which divergences from the norm have formed. It is clear that if our research had focused on just one site or location, it would have missed these complexities—if the research had taken place only at BBC Wales, we might have concluded that the culture of the BBC is one that is heavily focused on participatory journalism, whereas our other sites evinced a far more journalism-centred logic of production revealing the difficulty of challenging conventional hierarchies. The bigger picture suggested by our ethnography reveals that competing priorities and interests shape the use and views of "UGC," and that the term itself provides the grounds for contesting the meaning of journalistic work and identities, and the audience's place within these.

ENDNOTES

1   The BBC produces news for a variety of formats and media platforms, and also for diverse audience groups. For this reason we observed journalists at work on television, radio, and Internet news, as well as those producing for national (what the corporation terms "network") news audiences as well for local and regional audiences. Local news on the BBC is primarily delivered by its large network of local radio stations and its features-led BBC Local (formerly called Where I Live) websites. Regional news can be found on the larger regional and national news television programmes (e.g., Wales Today for Wales, and Look North for Yorkshire and other regions in the North of England) as well as dedicated regional "hard news" websites.

2   We also carried out a representative survey of the UK population and focus groups with 100 BBC news audience members to ascertain how many people have contributed audience material and what people think of increasing uses of amateur content, a content analysis of BBC news output at local, regional, and network level to find out how often the BBC calls for and uses different kinds of UGC, and a non-representative survey of 695 people who contribute material to the BBC news website. We have addressed the findings of our audience research elsewhere (Wardle, Williams and Wahl-Jorgensen, 2008; Wahl-Jorgensen, Williams, and Wardle, 2010; and Williams, Wahl-Jorgensen and Wardle, forthcoming).

3   Set up in 2005, the UGC Hub is at the centre of the BBC's response to the increase in citizen participation in the media. It consists of a team of 23 journalists based in the new multimedia newsroom in London and has grown significantly in size in recent years (at the time of our research there were 12 journalists at the Hub, and when it was set up it had a workforce of just three). Hub journalists are responsible for the smooth running of the Have Your Say news discussion website, for eliciting and processing audience material in a number of other ways, and for distributing audience material to news teams across the BBC. On an average day the Hub receives around 10,000–12,000 emails, as well as hundreds of pictures and video clips, sent in by the public. Editors at the hub envisage it as a centralised resource which can feed audience content to the corporation's outlying news operations, and hub journalists often placed there on secondment from different parts of the BBC in order to facilitate close working relationships with key departments and newsrooms. In practice, however, the hub's service is far from comprehensive, and it inevitably has closer and more effective relations with some newsrooms than it does with others. It operates as the most visible portal for the public to submit content to the BBC, but many newsrooms and individual news programmes have their own (often very effective) modes of eliciting audience material and so do not use the hub as extensively as others.

4 We describe BBC Wales as a "national newsroom" using BBC's terminology, insofar as the newsroom serves the nation of Wales. However, it should be distinguished from what the BBC describes as "network" sites, which serve the whole of the UK. In practice BBC Wales operates much like a regional hub.

CHAPTER TEN

Convergence in the News Wholesalers
Trends in International News Agencies

Chris Paterson

Trends in global news agencies play a crucial role in defining what is news in online journalism, as newsrooms heavily rely on wire material. Revisiting two decades of research on Reuters and AP, this chapter documents the hesitant strategies of the agencies towards convergence and the implications of digitization for news production. While multiskilling has not been favored, spatial proximity of text, photo and video producers in central newsrooms and correspondent bureaus is seen as advantageous to facilitate co-ordinated coverage of major stories. The role of editors has evolved from fostering the detection of original stories to managing the constant flux of news material, among which video has grown central over time.

EDITORS' NOTE

News agencies continue to be the dominant manufacturers of original information used by online journalists, and their product has consistently appeared in a barely reworked form in the output of leading online news sources (Paterson, 2007; Magid 1997). The shift over the past decade and a half to online news production has in fact been, for many news organisations, a shift away from original news production to a passive model (McManus, 1994; Lewis et al., 2008) of reproduction of the journalism created by a small number of sources.

For example, in the first volume of *Making Online News* Garcia quoted an Argentinean online journalist explaining his "second class" status in terms of print reporters knowing "that we do not work with sources, that we work with wire stories and ...we are 'clipping' wire texts." Convergence has meant for most newsrooms "fewer reporters, less real news gathering," as one reporter told Friend and Singer (2007). As Domingo concluded (also in the first volume of this book), "The daily routines of online journalists are more concerned with filtering and selecting a constant flux of information rather than an active search for stories." It has been the business of the wire services to provide that stream of "news" (broadly defined)—especially the "foreign news" element of most media—and online journalism's insatiable appetite and limited originality have served news agencies well.

This chapter examines convergent news production[1] at the two largest agencies, Thomson-Reuters and Associated Press.[2] There has been little research examining the practice of convergence within news agencies or news agency journalism in any form. While we cannot understand the totality of online journalism without ethnographic investigation of these organisations, to date, we have oddly ignored wholesale (Boyd-Barrett, 1980) sources of news. This account by one of the editors of this volume is unusually included despite only a limited amount (several days) of recent ethnographic research in order to highlight that lack of research with news agencies. My research consists of nearly two decades of occasional observation and interviewing with Reuters and Associated Press—particularly in their television news divisions in London.[3] This chapter should still be considered preliminary; a good deal more research within news agencies is needed to describe the convergence processes underway and the impacts of those process on journalism globally. As Quandt and Singer (2008) observe, convergence has often been seen by news organisations as an attractive business model in which more is produced with less: fewer people, less resources. Journalists, on the other hand, have been reluctant, sceptical, and even hostile to top-down approaches to convergence. But this short chapter will not attempt a detailed review of the myriad and always shifting conceptions of news convergence, since other chapters in this volume do so well.

News Agencies and Convergence

News agencies should be at the forefront of multimedia, converged, digital production, given their approximately 160-year histories in global news gathering and distribution, massive news processing apparatus, and a century of producing news for multiple media as efficiently as possible, and, like online news, doing so without deadlines. In the introduction to a recent anthology concerning national news agencies, Boyd-Barrett writes "it is not a surprise...that agencies are also leaders in the multi-media, multi-modal era of convergence" (2010, 27). Within the international agencies they have long had the sense of leading the way in convergence. An AP editor stated in 2002 they have "been far ahead of the convergence curve... because we've been sharing information across media for so long" (Bertini, 2003). There is, then, the paradox of a public perception (which they encourage) of leading the way in the 'convergence revolution', while in fact this revolution is a far more evolutionary and very long-term process of maximizing efficiencies in news production across a range of media channels. For example, the AP (established and funded by US newspapers), has been the main supplier of news to the US radio industry since the 1920s. Perhaps the most visible aspect of that evolution in the past two decades has been the costly embrace of video by both global wire services, Reuters and the AP.[4]

The agencies have separate divisions for text, photography, and video, which occasionally coordinate their work but more often act autonomously to distribute their product quickly. Speed drives everything in news agencies, and observation of rival agencies (and major broadcasters) determines what is fast enough. The contemporary news agencies are formed from separate specialised companies which evolved to deal exclusively with different forms and genres of news. And so it is perhaps unsurprising that distinct television agency, and to a lesser degree, photo agency, cultures remain within their operations despite considerable efforts toward integration.

Journalists in central newsrooms monitor the same sources: theirs and competing news wires, 24-hour broadcasters, and (at AP, though not as much at Reuters) Yahoo and other leading Web news portals (although, as I note below, to a lesser extent than might be expected). But the extent to which shared information has changed traditional—single media—production routines is unclear. It was apparent, for example, that video editors in both Reuters and the AP continue to work primarily with information and pictures gathered by their video journalists in the field and refer only infrequently to the text or photo product of their own agency. They occasionally check the stories they are putting out against what their text service is doing or (less frequently) what other media are doing, but the influence of those sources is not strong.

There is little research into the workings of news agency bureaus (each major agency boasts 200 or more bureaus, though only a small fraction of those are large operations contributing frequently to global news flow), but news managers speak of efforts to ensure convergence, and integration of media operations is happening both in the bureaus and in the central newsrooms. One Associated Press editor suggested that the AP's large Jerusalem bureau has been a "poster-child" for convergence in the news agency but mostly in terms of pooling information and resources among the different media sharing the bureau; he emphasised that there were no attempts to compel individual journalists to file reports *simultaneously* for different media.

Interestingly, the AP Television News' director of news told me the AP is encouraging journalists to train to do exactly that—unilaterally cover news stories for multiple media using the latest multifunction digital cameras. Upon qualification as a multimedia journalist these individuals might be assigned to cover lesser stories for each form of media *simultaneously*, providing text, photos, and perhaps audio and video. Embracing this widespread trend in the news industry will help news agencies produce more content across all 'platforms' at less cost, but proving or disproving their claims that news quality never suffers is difficult. He was quick to emphasize that this approach is used just on minor stories, and that such training is not mandated. Reuters' editors I spoke to, conversely, suggested such approaches hold little sway in their company at present, although Reuters has engaged in experiments with "multimedia journalism" (Kiss, 2007).

Multiskilling and Digitization

Because the researcher has had unusual access[5] to one leading news agency (Associated Press Television News) from its inception in 1994 (originally as APTV) to the current day, this chapter presents, in effect, a (brief) multi-decade longitudinal ethnography. Two interrelated trends which are often considered as significant constituent elements of newsroom convergence are the digitization of news production and trends toward multiskilling. There would be the risk of de-emphasizing the long evolution of these processes through research focussing only on the contemporary newsroom. The news agencies themselves contribute to such de-historization of their convergence processes through the modern hype they produce about their own convergence initiatives, publicity which obscures past efforts of a similar nature or the interplay of forces shaping news production which go beyond management decree.

Television news agency newsrooms and wire service newsrooms of the 1980s were highly specialized. In television, video editing, story writing, satellite coordination, news planning, and other tasks rarely overlapped. Through a steady demand for multiskilling imposed by the corporate hierarchy came the gradual erosion of specialized roles, with the objective of a more flexible work force capable of completing a wider range of tasks. The result of this process is that many news employees can move with little retraining in and out of a variety of newsroom posts but also that some specialized technical roles, especially that of video editor, have vanished (this was also a by-product of de-unionization). Through the 1980s, the international image of the television news agencies was very much manifest in the exacting and unique nature of the editing of video output, but with the de-professionalization and digitalization of this task, video is now edited quickly and in a fairly non-distinct fashion by some of the least senior and least experienced staff. Television news agency professional editors often possessed the ability to recognize in an instant key people in the news stories of the day and rapidly identify and cut into place the most valuable images from a massive amount of incoming video. As new or casual staff increasingly took on responsibility for selecting and editing images, that expertise and the knowledge of how to tell a televisual story to greatest effect may have been lost, though such change is difficult to quantify.

When I began to observe television news agencies around 1991, there was still the odd typewriter in evidence in newsrooms, and up until the mid-1990s the conversion to fully computer-based news production was occurring uneasily. The Associated Press began developing newsroom computing systems as a commercial venture from the 1980s, originally as "AP Newsdesk," and required its own newsrooms across its various media channels to adapt to the system—sometimes to the resentment of the journalists involved. Worldwide Television News journalists, in the years before that company was bought out by the AP, struggled with the Basys newsroom computer system, a system developed primarily for American local broadcast stations and initially not well suited to international television news agency production.[6]

Physical Integration

The variety of processes constituting convergence were well underway by the late 1990s, but the principal catalyst was the opportunity to merge under one roof operations for different media which had previously worked with geographic separation. For Reuters this happened in 2005 when their television news operations—based for many years on Greys Inn Road in London alongside the broadcaster ITN—and their print news operations based on Fleet Street—were merged together in a vast modern newsroom in a new building on Canary Wharf alongside London's leading international finance companies. With its acquisition of the oldest television news agency, Worldwide Television News (WTN), the AP took over the large canal-side former gin warehouse in the Camden borough of London, where WTN had operated for over a decade alongside other television companies. The AP created a new television newsroom there after the WTN acquisition and decided by 2007 to make the distinctive building the hub of their converged news operations in Europe (the bulk of AP operations are in their New York headquarters). A new multi-media newsroom in Camden opened in 2008.

The AP argues its move toward convergence has deeper roots. They promote their Washington, DC "Broadcast News Centre," dating to 1974 but substantially redeveloped in the 1990s and early 2000s, as "APs first multimedia newsroom," and in an internal report they detail the methods by which journalists from different media cooperated, emphasising the physical proximity in the newsroom of those different media and noting journalists in the large open, multimedia newsroom from "national, metro, and APTN are all within line of sight (and an easy shout) of each other" (Bertini, 2003). But contemporary conceptions of convergence in the international news agencies, while keeping the focus on physical proximity of key editors, have emphasized quiet, smooth workflow facilitated more by electronic messaging than shouting but with the ability of editors to move quickly to the desk of a colleague from another form of media to discuss a developing story or opportunity for integrated news coverage.

In London, both news agencies adopted similar physical arrangements for their converged newsrooms. In each, those news workers processing video—what once was known as the television news agency part of the company (a terminology now out of favour)—occupy the most central part of the newsroom, with typically larger text operations and the similarly sized photo operations adjoining at the sides. Television output desks are located within a few meters of those managing input, and regional editors for television are also nearby but a further distance from the prominent video walls displaying incoming video feeds, television output, and the output of selected major broadcasters. This is because regional editors are meant to be looking ahead rather than focusing on news of the moment.

With no direct access to the output of their rival news agencies, other means to monitor what the competition is doing are sometimes used. This may include, for example, AP editors checking a Reuters text story on Yahoo against what they have covered, television journalists from either agency watching what the other sends to the Eurovision television news exchange and keeping an eye open for breaking news coverage which isn't being originated by them on leading rolling news channels. At AP's new London newsroom any computer can look at "about 80 sources" coming into the building. But there is also now a remarkable confidence that between what their planning desks have put on the agenda for the day, and what their bureaus have called in, constant monitoring of other agencies or other media in the central newsrooms is unnecessary (despite the considerable presence of that other media).

The co-location of separate media operations is one of the most visible aspects of convergence in the newsroom, but its efficacy remains questionable. Bechmann (in this volume) examines newsroom convergence in terms of network theories (especially, Castells, 1996, and Latour, 2005) and observes that the focus by many news organisations on the *spatial* proximity of journalists seems peculiar, given the capability and reach of their existing news production networks via the various communications channels at their disposal (email, messaging systems, shared news production computer systems, etc.). She points to research suggesting proximity may facilitate the ease and frequency of communications between individuals, but also may lead to unneeded interruptions of work and (referencing Kraut *et al.*, 2002) may result in workers being excessively attentive to the 'local' at the expense of the 'remote.' The implications for news agency work—centred on gathering and processing information about the 'remote'—could be significant, but research on contemporary production practices in news agencies is still too limited to test this.

Collaboration between Media Teams

Various processes to combine the work of different media units are now standardized. In the last decade, the Associated Press seems to have moved from a philosophy of "everyone protects every service" (that is, each form of media output; Bertini, 2003) to striving for simultaneous delivery of incoming news to every platform, supplemented by discussion between editors of different media on larger stories. My research in the 1990s suggested that generally television news production (in Reuters and AP) was not well integrated with what the text side of the company was doing and that communication between the two was rare. Now, at the Associated Press, a morning editorial meeting between senior editors for text, photos, and television takes place to summarize coverage plans for the big stories of the day from across the world and for major stories anticipated later. But Reuters currently places less emphasis on systematically holding joint editorial meetings and sharing coverage plans. The AP's London director of news argues that the AP seeks "joined up journalism." He said "our mistaken belief was if everyone worked together in bureaus it would all be joined up [but] you got text, photo, video stories that mirrored each other." They have tried instead to create a process of deciding centrally what the story is and what are the best tools to tell the story, and then directing different media units to focus on telling different parts of the story. Using both a Top Stories Desk (Associated Press, 2008) focusing on planning multimedia coverage for larger stories and a new role of "cross-format co-ordinator" on major stories, the agency seeks to facilitate this "joined up" approach.

A final, and perhaps most highly publicised, form of converged journalism is increasingly being practiced at the AP. This is what they term "1-2-3 filing," whereby every new story entering their news production machine starts as (1) a short, about fifty-character, headline, which can be directed immediately to subscribing services such as onscreen news crawls, mobile telephone news updates, and online news providers—or, simply spark further investigation within the AP; then (2), a 130-word summary of the story, written in a "platform-agnostic style " that can be read by television presenters, used online, or used as a basis by media organisations for updates to their stories; and then (3) where a story requires more detail, a rapidly filed 500-word story.

Digital Video in News Agency Work

The very costly analog delivery of television images from remote locations around the world to television news agency London newsrooms was the core role, and a leading cost, for the news agencies from the early 1960s to the late 1990s. A few minutes of satellite time from some locations could cost thousands, or even tens of thousands, of dollars. One effect was to shut out many remote parts of the world—where only costly satellite delivery options were available—from the international news map (see Paterson, 2011). A number of technologies have now radically altered these traditional processes of image delivery, along with news agency work practices, and enabled reductions of newsgathering costs and the reallocation of resources. Nigel Baker, now an AP vice president but previously the managing director of APTN as it grew rapidly in the late 1990s, argues that the key moment where digital technology enabled video to surpass the written word in determining the nature of international news coverage was in 1997, when computers were used to compress video and transmit it across the world by way of satellite telephones (long used by the shipping industry but few others). Baker claimed (2009) that this technology allowed the AP to file daily television reports from Zaire (now Republic of Congo) as rebel forces proceeded to the capital to overthrow the government, while other news agencies were filing images up to three days old.

Video images are now transmitted to central newsrooms—either London or a few regional centres—by any of several digital methods. The decision of which to use is based on cost, availability of technologies, and the quality of image desired. Image quality may sometimes be sacrificed for the sake of speed or cost. The preferred method at the television news agencies now is FTP[7] delivery, where a digital file directly from a camera or from the camera operators' laptop computer is sent via the Internet to the newsroom. This method depends on access to a broadband internet connection. If none is available, alternatives will be used. Live video that news agencies often strive to provide for larger stories requires other technical arrangements; in these cases, good video can still only be obtained from costly satellite links; digital satellite news gathering has, however, lowered the cost and simplified the process (for more on the agency role in live television, see Paterson, 2010).

In the newsroom, the ability of any news worker to access at their desktop a video file as soon as it is transmitted from the field has had a number of effects. It influences editorial convergence, increasing the impact of the television side of newsgathering on the text and pictures side of the business. Traditionally, television photographers get to and transmit the story first, but the rest of the news agency will wait for a still photographer and a text reporter to provide those aspects of the story—thereby increasing the amount of information gathered on any breaking story and allowing a bit more time for analysis of a developing story. But there is an expectation now that photo journalists will take a frame from incoming video if they don't have photos, and distribute that quickly. While it won't be of the same quality or tell a story as effectively as one provided by a still photographer, it may be the photo which gets the most exposure because it is transmitted first. Text writers will be expected to transmit information based on what they see in the incoming television pictures. Whether or not this comes at the cost of developing their own story more thoroughly has not been subject to independent research. News agencies tend to celebrate the availability of more information at the desks of their journalists, but little is said about what reporting processes are sacrificed as journalists deal with new forms of information.

Original Reporting vs Repurposing of Content

Bardoel and Deuze (2001) suggested that in new forms of "networked journalism" a new focus on "windowing" of content (following Fulton, 1996)—packaging it for various output channels—has placed a renewed and welcome emphasis on journalism's core functions of gathering and disseminating information: journalism's "great opportunity in keeping up old values in new media." But Currah (2009, 3–4) draws a compellingly different picture in his recent overview of the UK media, based on extensive interviews across mainstream media producers. He argues that "many publishers are morphing into a form that favours the processing rather than the generation of content...there is a clear danger of a sharp reduction of spending on original newsgathering."

My limited observation in recent years in the international news agencies suggests an ongoing struggle to maintain a balance between these competing forces. For the traditionally key gatekeeping positions in the international newsgathering machine, such as the television "editor of the day," (EOD) there is tangible shift over the past two decades from a role of seeking to manage and monitor unfolding events around the world and interact directly with the staff "on the ground" reporting on them to a less hectic and more managerial role of ensuring the flow of pictures from planned events and monitoring other parts of the organisation and other media to minimise the risk of missed stories or missed opportunities for internal collaboration.

While it is an equally stressful role, it places far more trust in "the machine" (the agency bureau structure, the planning desk, the monitoring of allied and competing mainstream media channels) to discover news and send it in for its initial processing and packaging to feed an ever-increasing array of outputs. This is not a diffusion or disintegration of the gatekeeper: the EOD and other senior central newsroom staff in the news agencies still maintain a strong gatekeeping role in line with longstanding news agency tradition (i.e., Paterson, 1997). In seconds—at

times—they sketch out or utterly reject news coverage of events as they hear briefly
from editors in the regions around the world and review the day's news diary. Stories
that are rejected may or may not become local news—but they *will not be global
news* and will have limited historical significance as a result of those decisions.
When those decisions have the potential to influence public policy, i.e., international
aid or military action, their impact might be measured in thousands of lives.

Conclusion

News agency central staffers take their roles no less seriously than they did twenty
years ago, and the pressure to be the first to produce pictures from an important
breaking story is no less apparent. The news agencies might well argue that it is
both the reduced dependence on satellite news delivery, with its high cost and poor
reliability as well as their success in devolving editorial power and newsgathering
resources to regional news production centres and to key *bureaus* that has made
the newsflow through London headquarters less chaotic (the days of editors scream-
ing into three telephones simultaneously seem long gone). That improved efficiency
may well give news agency journalists more time to conceive and produce original,
enterprise journalism or to break news other media have failed to discover.[8]

But the picture that emerges from this limited observation—within the broader
context of observing two decades of newsroom change—is of an ever-shrinking
diet of daily news largely defined by what other media are doing; the acceptance
of a role of trickling new facts, images, and sounds into the global swirl of "news,"
the contradictory impulses to be "first" with those while rarely striking out into
coverage of controversial or under-reported stories, an ever-increasing confidence
that the organisation knows what is important in the world, and a reduced sense
that their primary function is to write human history—with all the responsibility
that entails. There is a pressing need for new ethnographic research into the vast
news production processes of the news agencies—the services originating the news
that defines online journalism.

ENDNOTES

1   In this chapter, I use the term 'convergence' primarily as an indicator of change in micro-level
    news production although its application in the news agency context ranges across the breadth
    of meanings associated with the term, from the integration of previously distinct media outputs,
    to changes in news gathering, to industrial realignments.

2   While they are not a focus here, authors in Boyd-Barrett (2010) detail the role of the national
    news agencies. Also see the case of Malaysia in the chapter by Firdaus in this volume.

3   Reuters took over the television news agency Visnews (which it had co-owned with broadcast-
    ers) in 1992. In 1994 the Associated Press entered the video business with APTV, which

transitioned in 1998 to Associated Press Television News (APTN) with the buyout of WTN. WTN had been the focus of this author's extended ethnographic research in the mid-1990s, but weeks of observation and dozens of hours of interviews were also compiled from Visnews, Reuters, and APTV in this period. Supplementing my own observations and interviews are descriptions of the AP's New York and Washington operations by Edmonds (2006) and Bertini (2003), respectively, and the public pronouncements of the agencies themselves. A production research project focussing on convergence at international news agencies now underway is that of Jones (Jones and Paterson, 2010), and the author is grateful to her for suggestions on this chapter.

4 The AP purchased the WTN television news agency for about 50 million US dollars in 1997 soon after spending many tens of millions establishing their own television news agency from scratch.

5 The only other substantial ethnographic project with a news agency central newsroom was also thanks to access granted by the Associated Press (Clare, 1998).

6 Basys was taken over in the mid-1980s by WTN co-owner, Independent Television News, and then by AVID (my observation, also Bertini, 2003; Luff, 2002)

7 File Transfer Protocol, one of the oldest technologies of the Internet designed to transmit files from a computer to a server.

8 I am not aware of any comparative content study of news agency output—pre and post-convergence—to demonstrate whether or not this is the case.

CHAPTER ELEVEN

A News Portal without a News Team
Journalistic and Marketing Logics
at the Malaysian National News Agency

Amira Firdaus

The Internet has offered national news agencies the opportunity to extend the reach of their services to non-media customers. This chapter analyzes the case of BERNAMA, highlighting the tensions between journalism and marketing in the process of blurring the traditional definitions of news agency with their online services. While the journalists defend their role as a service to promote Malaysia's place in the world, which fosters a pro-government bias, the marketing department in charge of the website strives for neutrality in order to avoid undermining their business. The multifaceted implications of strategic decisions in the design of online news products are discussed, with commodification of news being one of the most worrying for the author.

EDITORS' NOTE

Much of the existing literature tends to be historical or systemic accounts of global news agencies, documenting the evolution of major international news agencies and evaluating the news agency role within different political systems (e.g., Boyd-Barrett & Rantanen, 1998). A recurrent theme in news agency literature is that of 'news agencies in crisis' (Boyd-Barrett, 2010). As Boyd-Barrett and Rantanen (2004) explain, the crisis of news agencies is 'practical' and 'conceptual' in nature. Practically, news agencies face competition in crowded marketplaces brought on by increasing costs of news production and proliferation of news outlets facilitated by developments in broadband, mobile, and satellite technologies. Conceptually, traditional definitions of news agencies are getting more difficult to maintain as news agencies expand their activities to go beyond the wholesaling of news to include direct 'retail' sale of news and information via networked technologies.

However, as Boyd-Barrett notes in the opening to a volume entitled *News Agencies in the Turbulent Era of the Internet*, "the Internet had heightened the visibility of news agencies in the attention of news consumers and this added considerable value to efforts to develop positive branding of news agencies...some were beginning to deliver services better suited to a more commercial, multimedia environment." (Boyd-Barrett, 2010, p. 10). Such "entrepreneurial" efforts are designed to bring "innovation, diversification and market segmentation or specialization" into the news agency business model (p. 23). BERNAMA, the Malaysian National News Agency, presents one such example of news agency entrepreneurship, with its active marketing of news feeds and subscription-based mobile and online products to organizational clients (e.g., media, government agencies, financial institutions) and to individual news consumers. From the divergent vantage points of marketing executives and journalists, this chapter aims to join the current discussion by illuminating the disconnect between journalistic and marketing logics in a specific case where the selling of online news products is divorced from the journalistic production of news.

BERNAMA and the Internet

Headquartered in Kuala Lumpur, the Malaysian National News Agency, BERNAMA, maintains an extensive domestic network of bureaus, and a small number of foreign bureaus and correspondents. BERNAMA's Editorial Department oversees its three main News Services, namely General/Domestic, Financial, and Foreign/International. The General/Domestic and Financial News Services maintain both English and Malay Sub-Editing Desks, and a separate Mandarin News Desk. The Foreign/International News Service maintains another separate Arab News Desk.[1]

BERNAMA's news portal consists of four separate sites: a Malay language site, an English site, a Mandarin site, and an Arabic site. The two main English and Malay sites share a common layout and almost similar news line up, whilst the Mandarin and Arabic sites each feature different layouts. The main Malay and English sites include frontpage links to a number of BERNAMA's subscription-based products and services, including mobile news, numerous domestic and international news services, and several 'microsites.' Microsites feature news and events related to clients from major industries (i.e., Banking and Finance, Auto, Aviation, Maritime, Education and Halal). In addition, there are also links to non-news services such as photo archives, research databases, PR services, and information systems services.

The portal's news content consists of stories filed by wire journalists into BERNAMA's internal news management system, the BESSAR[2] Editorial System Terminal (BEST). A team of around eight staff with marketing backgrounds, headed by a website editor, select stories from the BEST system to subedit and upload to the website, without further input from wire journalists or editors. This website team currently sits in the Editorial Department, although it was, until recently, part of the Media, Portal and Photo Marketing Division. While the website team is responsible for news content, the marketing division oversees the portal's advertising content and its subscription products, including microsite content.

Methodological Challenges

This study began with the intention of conducting an ethnographic study of the online newsroom of BERNAMA. It was guided by the naïve assumption that BERNAMA.com had its own dedicated team of journalists, as do its TV and radio counterparts (BERNAMA-TV and BERNAMA Radio24). In reality, it doesn't.

Adding to the methodological quandary of conducting an ethnographic study of non-existent online journalists was the unexpected setback of being refused permission to speak to website staff. From an ethnographic standpoint, these individuals' lived experience cannot be substituted by those not directly involved with the daily work of selecting and subediting stories for the website. However, from a more general methodological standpoint, difficulties in gaining access to organizational members often means that researchers must be flexible in their selection of research participants. As Sullivan (2004) suggests in a paper discussing strategies for in-depth interviews with media workers: "[M]ost researchers can't afford to be terribly systematic about sampling [media] elites. Outside researchers generally have to take what they can get, and should be opportunistic about getting access to media companies and their employees." (p. 13). Given that the website's news stories are largely finished products long before each story is selected for upload, in-depth interviews with staff from BERNAMA's other departments were found to be useful for gleaning insight into the journalistic and marketing logics shaping news content available on the website. Accordingly, this chapter is based on interviews with journalists from the Editorial Department, as well as staff from the Media, Portal, and Photo Marketing Division of BERNAMA's Business Department. All interviews were conducted face to face at BERNAMA's headquarters in February and August 2010.

BERNAMA's online offerings are largely shaped by two intertwining influences: a pro-government development journalism philosophy that characterizes much of Malaysian mainstream media and an underlying marketing logic that is driven by an organizational desire to capture market share in an increasingly competitive marketplace.

Journalistic Logic

As can be gleaned from its online news portal-with its link to dedicated pages on government initiatives, its microsites highlighting major national economic sectors, and its general focus on government news, BERNAMA's organizational identity is indeed very much that of a government agency, even though it is technically a statutory body. BERNAMA is generally expected to place national interest as a priority, much in the same way that major news outlets in the 19th century (i.e., Reuters, Havas, Wolff/Continental) reflected the national interests of their respective governments (Boyd-Barrett, 2008/1998). Typical of mainstream media in Malaysia, BERNAMA's organizational identity *vis-à-vis* government can be traced to the pro-government school of development journalism which emphasizes the role of the press in helping government to promote and preserve economic prosperity and political stability, and a "responsible exercise" of press freedom to maintain economic and political cohesion (see Xu, 2009, pp. 361–363). In Malaysia, "constructive partnership between the press and government is highly expected in theory and respected in practice" (Xu, 2009, p. 359). One example of 'constructive partnership' is the Arabic version of BERNAMA.com, part of BERNAMA's strategy to expand into the Arab-speaking world:

> This started about five years ago when we wanted to encourage more tourists from the Middle East to come to Malaysia and to invest. So that's why we started [the Arabic News Service] to support the government focus on Middle East countries, to tap opportunities, trying to lure them to come to Malaysia…[Audiences] in the Middle East normally follow our website because of the news that they get [about Malaysia]. (deputy editor-in-chief, Foreign/International News Service)

The deputy editor-in-chief of foreign news further adds:

> [I]f Malaysia announced a policy, maybe to encourage FDI [foreign direct investment]… that kind of stories we put up…Even 1Malaysia,[3] we [would] normally put up. Because we just want to share [news] about 1Malaysia [with] the world. (deputy editor-in-chief, Foreign/International News Service).

As noted by Xin Xin in her work on Xinhua, the Chinese national news agency (2010), agencies play a role in exercizing "soft power" and public diplomacy. Boyd-Barrett further suggests, "This may occur, for example, where an agency is deliberately guided by government towards the construction of a positive perception of its nation, among domestic, regional and international audiences" (2010, pp. 16–17). In the case of BERNAMA, it is likely that national development goals are ingrained within the organizational culture and echoes what researchers have concluded elsewhere about post-Soviet Russian media: "ideologically they remained national services, which thought it their duty to meet the interests of the state by shaping the country's positive image." (cited in Vartanova & Frolova, 2010, p. 229)

To be fair, it can also be argued that as a *national* news agency it is simply natural that BERNAMA would focus on national issues and government policies. As the deputy editor-in-chief of foreign news points out:

> Although [critics] say "it's [all] official [news], it's coming from the government," but *someone* [interviewee's emphasis] has to come out with [those] stories!

The partnership between BERNAMA and government facilitates access to government sources and makes it easy for journalists to function as 'disseminators' of information gathered from official sources rather than fulfilling 'adversarial' watchdog functions or 'interpretive' analysis roles that would often necessitate more distance, rather than closeness, between press and government (see Weaver & Wilhoit, 1991, for roles of journalists).

Somewhat like China's state-owned news sites (see Lagerkvist, 2008), news stories on BERNAMA.com tend to be supportive of macro-level government policies and ministers, reserving criticism for meso-level performance of government departments (e.g., tactical blunders, oversight) and micro-level lapses of government officials (e.g., corruption, conflicts of interests).

Marketing Logic

As news agencies increasingly venture into online news via a transnational Internet, international news exchange agreements—once the key for news agency survival—are no longer able to protect news agencies from competition. Like many 'Internet-era' news agencies, BERNAMA's 'entrepreneurial innovations' address new, transnational forms of competition by utilizing networked technologies to capture new markets. Through its various portals, BERNAMA engages in cross-promotion and distributes and tailors its agency news products to both media clients and to individual consumers. It does so by taking full advantage of its competitive strengths, namely, its reputation as a credible source of Malaysian government news and its partnerships with foreign news agencies like Reuters and EFE:

> BERNAMA's strength is still in governmental news. That's why we have not lost any subscribers. Although in terms of financial strength, and ability to report, newspapers… can employ more reporters than BERNAMA. Where they find BERNAMA still relevant

is BERNAMA is the authoritative source of government news. (retired editor-in-chief & CEO)

Leveraging on its full market control of domestic agency news, BERNAMA services both domestic as well as foreign media clients by offering its own news content and by redistributing the content of other national, regional and global news agencies. It retains rights to redistribute foreign news feeds by delivering them to media clients that would otherwise be prevented by cost or by simple lack of interest. BERNAMA currently controls Malaysia's domestic market for foreign news by offering clients attractive pricing, free access to certain feeds, and customized foreign content, including sourcing for content not already included in its packages. Thus, despite its foray into 'retail' news via freely available online news and other subscription products available to individual news consumers, BERNAMA retains its commitments as a traditional news agency through providing 'wholesale' news to other 'retail' news outlets. At the same time, the freely available daily news on BERNAMA's website indicates that the news agency may also be competing with its own subscription products by disseminating news free to the public. One strategy to mitigate such competition is to restrict free online access to just a portion of available copy:

> We have a policy of [uploading only] 60% of the news....Mainly because of our other products, other than the website. Such as mobile, BlackBerry, and news subscribed by other people who might want to redistribute the news. So we don't put out everything we have on the website, otherwise we might kill the other businesses! (sub-editor, Media, Portal and Photo Marketing Division)

In addition to its 'wholesale' wire business, and its 'retail' online and mobile business, BERNAMA also routinely "performs functions not dissimilar to those of government press or information office," as do a number of state-supported news agencies elsewhere (Boyd-Barrett, 2010, p. 11). Sub-editors from the marketing division and the website team, as well as wire journalists from the editorial department, regularly process press releases and cover events held by other government agencies. This practice is also extended to events organized by BERNAMA's non-media clients for whom promotional pages are maintained under BERNAMA. com's industry-based microsites. Coverage of organizational clients' events is part of BERNAMA's marketing logic, and the existence of online products and services facilitates the organization's relationship-building strategies with important clients. Even if a client event is not deemed newsworthy by 'retail' news outlets subscribing to BERNAMA (or even by BERNAMA's wire editors), clients are assured that their news will at least make it onto BERNAMA.com.

Tensions between Journalistic and Marketing Logics

As a *national* news agency, BERNAMA's journalistic logic is very much influenced by a pro-government development journalism philosophy as illustrated above. From a business perspective however, there is less commercial impetus to be pro-government than there is to distance itself from government.

Indicative of BERNAMA's efforts at distancing itself from government is its willingness to allow news critical of the Malaysian government to freely flow to its media clients (though perhaps less often to individual news consumers).

> No censorship [in our wires]…for example some Reuters stories criticize the Malaysian government…All the stories are just sent out as it is. So that's why we have subscribers like *Malaysiakini*.[4] A lot of the opposition [political parties]…subscribe [to] our news on their website, and also our mobile services. (media manager, Media, Portal and Photo Marketing)

Behind the decision not to self-censor its wire services is a commercial imperative largely at odds with BERNAMA's pro-government development journalism logic:

> And if we do take a pro-government stance, then we will lose clients, and also the foreign press will dismiss us as just a mouthpiece and so they will [not subscribe to BERNAMA]. So then the government [would be] unable to reach…the international and the local and all, because nobody will subscribe to our wire…I think the [government] understands that. (media manager, Media, Portal and Photo Marketing Division)

Notwithstanding the commercial need to distance itself from government, there also appears to be an internal perception that BERNAMA covers news in a somewhat 'neutral' manner:

> Our [news] in many ways [is] very neutral….Compared to other media…I mean, considering that we are [a] government [agency], I think that we are more neutral in many ways. (senior manager, Media, Portal, and Photo Marketing Division)

Most Malaysian mainstream news outlets are owned by government political parties whilst online alternative news sites and blogs are either affiliated with—or are strong supporters of—opposition parties. Thus partisanship is very much a hallmark of Malaysian journalism with news outlets openly aligning themselves with particular political parties.[5] As exemplified by the above quote, BERNAMA's 'neutrality' or objectivity is relative to the partisanship of other news outlets, and always in relation to its status as a 'government agency.'

In this context of media partisanship, BERNAMA's aspirations of journalistic neutrality may be one attempt to live up to the spirit of BERNAMA's establishment. Although BERNAMA is widely identified as a government agency, technically, it is an independent statutory body:

> I think because…BERNAMA was set up by an Act of Parliament. Our mission, vision, and all that…it's not pro-government.…Actually BERNAMA is not under the government, its under its own Parliament Act, but the government gives grants so that BERNAMA [will carry government news]. But that doesn't mean that we do not carry critical news of the government. Because sometimes [the stories] are merely reporting [about] incidents…[The government] appreciates that, because they would like to respond to [the incidents]. (media manager, Media, Portal and Photo Marketing)

Like its neutrality aspirations, however, this sense of independence is juxtaposed against BERNAMA's own internal perceptions regarding government reaction to criticism, even when the government is displaying an openness to criticism. To return to the earlier discussion of BERNAMA's journalistic logics, BERNAMA's retired editor-in-chief further explains government acceptance of openness in news coverage:

> It's good for the government that we are also seen to be more open in reporting…I think the media reflects the credibility of the government. The media of the country, if it's not credible, [then] the government is not credible. There is quite close linkages between the credibility of the government and the credibility of the media. (retired editor-in-chief & CEO).

Conclusion

From an ethnographic standpoint, tensions between BERNAMA's journalistic and marketing logics arise from a disconnect between the editorial department's development journalism pursuits and the marketing division's concern with capturing market share through a variety of networked delivery platforms. This disconnect is exemplified by the minimal interaction between journalists and marketing staff in charge of BERNAMA's online portal, and the lack of input journalists have upon their stories after filing them:

> … technology makes it that [the news] goes here and there.…They [the journalists] input the story, it goes to a system.…They don't have to interact with [portal staff]. (senior manager, Media, Portal and Photo Marketing Division)

Thus, unlike dedicated online news ventures, or even convergent newsrooms where online news is produced through dedicated efforts in online journalism, 'online news' at BERNAMA is more aptly described as 'online *news products*'.

Indicative of the marketing logic driving BERNAMA's products and services, much of their content requires paid subscriptions and suggests a view of news and information as a "commodity" (see Boyd-Barrett & Rantanen, 1998; Palmer, 1998; Winseck & Pike, 2007). As Michael Schudson (2003) notes in a chapter on news in the marketplace, it has become a general trend for news outlets to "[develop] projects or even whole sections for their financial potential, not for their news value"

(p. 122). Thus, commoditization of news may be a strategy for remaining relevant in the face of competition:

> If and when…the government feels that news agencies are no longer relevant, then that will be the day when BERNAMA will have to cease operations, because BERNAMA… [depends on funding] from the government. But at the same time, we are marketing our information products, to [generate income]. We [partly fund] our operations through our own commercial operations. (retired editor-in-chief & CEO)

BERNAMA's Media, Portal and Photo Marketing Division's earnings are estimated to amount to half of BERNAMA's income, and interviewees expected them to rise. If this trend is indeed indicative of greater financial independence for BERNAMA, the earlier mentioned 'news agency crisis' may be more a conceptual question of which of its identities truly reflects its activities: a national agency or a media business?

BERNAMA's role goes beyond the traditional news agency role of supplying news content to media clients to include media work for non-media organizational clients (e.g., promotional pages in BERNAMA.com microsites) whilst also engaging in direct retailing of news products to end consumers via online and mobile news services. BERNAMA leverages on its position as a 'constructive partner' to the government in the pursuit of government-defined national goals to carve a niche in government news, which it then aggressively markets alongside other online entrepreneurial ventures. This commercialization of news products is facilitated by digital technologies to deliver not only real time news but also online media services such as news coverage of client events.

To the extent that networked online and mobile news content is shaped by a combination of marketing and journalistic logics and to the extent that some news organizations treat online news as mere extensions of existing dissemination platforms and not as a distinct practice of journalism requiring dedicated online newsrooms and journalists, a new question arises regarding the extent to which online news—or rather, online news products—truly benefit the news organization and its audiences. In the case of BERNAMA, the lack of online journalism practices could be interpreted as rational cost-savings but may also just be an organizational oversight that simply widens the gap between editorial news values (i.e., the pursuit of pro-government development journalism) and marketing values (i.e., an increase in wholesale and retail news market share). Conversely, the lack of an online news team socialized to a pro-government development journalism ethos, coupled with the fact that the portal team is trained to produce news that captures the attention of online consumers (and by extension, online market share), may serve to balance the editorial department's pro-government slant with the more market-driven (and by extension more libertarian) business goals, thus resulting in a sum of parts that is a semblance of balanced-reporting.

ENDNOTES

1   Malaysia's national language is Malay, mother tongue of its indigenous population. English is widely spoken, particularly in business. Mandarin is widely spoken, particularly by Malaysia's Chinese business community. Malaysia maintains close ties with the Muslim world and many Arab countries, and maintains an Arabic service for clients and partners from Arab countries.

2   BESSAR is the acronym for BERNAMA System and Solutions Advisor, a wholly owned subsidiary providing commercial information systems services.

3   1Malaysia is a large-scale government initiative to foster multicultural unity amongst the country's different races. It should be noted that BERNAMA.com compiles 1Malaysia news stories in a special page under a dedicated link on the website.

4   Malaysiakini is an online news site often critical of the government.

5   Malaysia is a parliamentary democracy in which multi-party general elections are held at least every five years. The current government coalition (Barisan Nasional) has been elected into power in every general election since the country gained independence from colonial British rule in 1957. However, in the latest general elections in 2008, with the support of political bloggers and online news sites, a coalition of opposition parties gained control of several key states.

CHAPTER TWELVE

Blowing up the Newsroom
Ethnography in an Age of Distributed Journalism

C. W. Anderson

This chapter advocates a shift in the focus of online journalism ethnography, to put online newsrooms in the context of online public communication and the social networks that generate the news. It analyzes the dynamic relationship between professional media, bloggers and citizen journalism in Philadelphia and their sometimes diverging and sometimes blurring reporting practices and journalistic identities. The author presents Social Network Analysis and Actor-Network Theory as fruitful frameworks to reinterpret the centrality of professional newsrooms in local news ecosystems and illustrates the use of these theories with examples taken from a three-year ethnography.

EDITORS' NOTE

We stand at the threshold of a second "golden age" in news ethnography (Cottle, 2007). The chapters in this book and the previous volume(Paterson & Domingo, 2008), along with other recently published work (Boczkowski, 2007; Domingo, 2008b; Hemmingway, 2005; Klinenberg, 2005; Lewis, Kaufhold, & Lasorsa, 2010) and other research in various stages of production, provide happy evidence of the fact that scholars are rising to the challenge posed by the empirical changes in the journalistic field over the past decade and a half. There are many ways to understand the technological, cultural, and economic transformations of journalism—content analysis, large-scale data collection and quantitative analysis, surveys, social network analysis, critical theory, and many other methods. Ethnographic research is reassuming its pride of place on the list.

Ironically, however, just as ethnographic research is reemerging as a preferred method for studying newsrooms, the very walls of that newsroom, both physical and literal, are shifting dramatically. The developments that make newsrooms promising sites of fieldwork—the key roles played by boundary-spanning devices and information flows, the professional challenges posed to traditional journalists by amateur bloggers and new news startups, and an economic crisis in journalistic business models that is driving more and more newswork outside formal organizations—make them problematic sites of fieldwork as well. In short, the classic "newsrooms" of Gans (1979), Tuchman (1978), and others (Eliasoph, 1997; Epstein, 1973; Fishman, 1980) cannot serve as our only model for fieldwork in a world where the very definition of journalism is being contested on a daily basis.

In the pages that follow, I want to suggest a model of research that holds onto what is valuable in most newsroom fieldwork, but also opens both newsrooms and our imaginations to the larger journalistic "ecosystem" that is emerging and coalescing outside these newsroom walls. While the title of this chapter is deliberately provocative, I do not want to suggest that the newsroom has no role in contemporary journalistic production. Indeed, the research upon which this chapter is based concludes just the opposite—the newsroom's role remains central. However, we can no longer take its primacy for granted; the status of the traditional institutional newsroom must be continually problematized. I attempt that problematization in the pages that follow by drawing on a methodology of "networks" in two senses. First, I discuss my application of Howard's (2002) notions of the "network ethnography" to the world of journalism and to the mapping of journalistic ecosystems. Second, I discuss the ways that Actor-Network Theory can help us follow various journalistic "actants"—stories, facts, links, employees, technologies, work routines—across this suddenly wide-open communicative space. The remainder of my chapter discusses how these particular approaches helped me to glean unique insights into the shifting nature of newswork and journalistic authority in the 21st century.

Visualizing the Emerging News Ecosystem Through Social and Actor Networks

In *Breaking Journalism Down* (Anderson, 2009), my ethnographic and historical research into shifting visions and practices of local journalism in Philadelphia, the goal was to determine how the production routines of journalism, the organizational structure of journalism, and the authority of journalism were changing as they intersected with economic crises and digital newsgathering techniques. All three of these rather disparate interests were united in a focus on the actualities of news-work and the day-to-day practices by which journalistic identity was affirmed. To investigate these questions my research methodology, first and foremost, was ethnographic, and the study summarized within these pages is based on a long-term period of participant observation in Philadelphia. I spent three years engaged in the qualitative research of news in Philadelphia, with a period of extensive ethno-graphic immersion that lasted from March until August 2008. This immersive period was complemented by a period of follow-up research and secondary site visits from the autumn of 2008 until 2010. During that time, I conducted over 300 hours of observation along with more than 60 semi-structured interviews with journalists, editors, activists, bloggers, and media executives. Interviews and eth-nographic notes were then coded manually, following the general model of "grounded theory," with early observations used to determine initial categories, then iteratively revised after several cycles of additional fieldwork and coding.

One major difference between the data presented here and earlier sociology of news studies is that my ethnographic study analyzed an entire local news ecosystem as opposed to the study of a few key, usually traditional and large-scale, journalistic institutions. Since the cultural, economic, and technological challenges to journal-ism were both extra- as well as intra-institutional, I resolved early on that I could not simply study the *Philadelphia Inquirer* or the major area news website Philly. com. I wanted to study the entire "Philadelphia news ecosystem." As news produc-tion decentralizes, traditional methods of exploring the behavior of journalists "at work" grow ever more problematic. This is not an argument to the effect that examining journalists at work is methodologically meaningless; rather, it merely points out that the question of where journalistic work occurs is a difficult one. If this has always been true for journalism, it is now doubly so, as the Internet and assorted digital technologies flatten and disperse the (post)modern workspace.

To overcome these difficulties, sociologist Phillip Howard (2002) argues that two strands of research—the traditional ethnography and Social Network Analysis (SNA)—can be combined as a *network ethnography* in order to analyze selected "communities of practice." SNA "is the mapping and measuring of relationships and flows between people, groups, organizations, animals, computers or other information/knowledge processing entities. The nodes in the network are the people and groups while the links show relationships or flows between the nodes. SNA provides both a visual and a mathematical analysis of human relationships" (Krebs, 2008). This analysis ignores the traditional sociological focus on self-defined, close-knit groups and concentrates instead on links—on a node's "centrality," a network's periphery, "bridges," "clusters," "connectors," and so on. Often, the strongest ties of various network nodes span boundaries between apparently separate groups; in

Howard's study of the e-politics community, for instance, a strong link emerges between political consultants and open-source technology activists, a connection that might have been ignored through an exclusive focus on either one group or the other.

At the same time, Howard criticizes the "un-grounded" nature of Social Network Analysis. While almost no researcher using SNA adopts it as his only methodological tool, the fairly unsystematic follow-up interviews or questionnaires submitted to key network nodes often fail to provide the kind of deep, rich empirical detail afforded by ethnographic study. For this reason, Howard advocates the network ethnography to analyze on- and offline organizations The network ethnography uses ethnographic field methods to analyze field sites chosen via social network analysis. "Active or passive observation," writes Howard, "extended immersion, or in-depth interviews are conducted at multiple sites or with interesting subgroups that have been purposively sampled after comparison through social network analysis." (Howard, 2002, p. 562). A key step in formulating my network ethnography of Philadelphia journalism was thus the creation of social network maps of the online Philadelphia media sphere, which themselves drew on data compiled through both Issuecrawler and Morningside Analytics software.

It is important to note, at this point, that these network maps functioned only as a general guide and starting point for my research. To avoid the kind of schematization and structuration that would result from an over-dependence on SNA, I wanted to focus not only on the actors and institutions in the Philadelphia media ecosystem but also on how these actors interacted and networked across that ecosystem. In that vein, I adopted a *second* methodology of networks, called Actor-Network Theory, which is more focused on action and assemblage than on structural mapping. This perspective analyzes the relationships between the objects, people, and institutions within the larger news ecosystem. How did traditional news organizations interact with local bloggers, for example, and how did the actual stories produced by these different types of news actors themselves collide with and shape each other? Actor-Network Theory argues that social categories (whether we mean science, journalism or anything else) should not be approached as obdurate macrostructures; rather, they should be seen as the contingent assemblage of networks. ANT emphasizes the radical heterogeneity of both humans and non-humans in the makeup of these social assemblages; this argument—that "objects have agency"— is perhaps the best-known feature of ANT. Finally, ANT contends that the creation of "social solidity" must be seen as a process of *black boxing*, "weaving together human and non-human actors into relatively stable network nodes, or 'black boxes'" (Bowker, 2007).

Most of the research utilizing Actor-Network Theory for the study of journalism has focused on the manner in which ANT can serve as a theoretical framework for mapping journalistic innovation. Schmitz Weiss and Domingo (2010) explicitly identify ANT as a tool for modeling innovation, while Hemmingway (2004, 2008) and Plesner (2009) have used ANT in fieldwork on the adoption of new technologies in newsrooms. Turner (2005) has made a more general call for adopting an ANT perspective in newsroom research. Boczkowski (2010, 2004a, 2004b) and Boczkowski and Lievrouw (2008), while not explicitly drawing on Actor-Network Theory *per se*, have examined journalism from the perspective of science and technology studies, which share many of ANT's general perspectives on the relationship between technology, organizations, and culture. In addition to these studies of newsroom innovation and technological adoption, Actor-Network Theory can also be used to study the entire ensemble of practices by which technological devices, human actors, journalists, sources, documents, and hyperlinks are "networked" together to create both news products and news organizations. This is how I have used it here.

In sum, I've tried to combine Social Network Analysis and an ANT methodology to understand a local journalistic ecosystem through ethnographic fieldwork. Social network mapping provided a baseline structure. Ethnographic fieldwork filled in that structure with rich, thick details, helping us go beyond our initial SNA-identified targets and investigate other spaces and places in the news ecosystem. Actor-Network Theory, finally, focused on the way that people, objects, news stories, and news technologies interacted as they created relational webs of meaning across ethnographic space.

All of these theories are useless, however, if they do not help answer research questions. What did I learn during my lengthy ethnographic fieldwork in Philadelphia, and in particular, how did my methodological perspective facilitate these discoveries? Did I learn new things by blowing up the newsroom and focusing on the news ecosystem instead? I now want to detail some of my research findings, discussing the way that these findings might help us carry the ethnographic tradition into a new journalistic age. For the purposes of this chapter, I want to detail three important points: the way that the digital "presence" of bloggers and citizen journalists affected the role conceptions of reporters, the diffusion and institutional re-appropriation of particular forms of newswork (particularly aggregation and reporting), and finally, insights into the manner in which news stories diffused in this new journalistic universe. These findings, I argue, were at least partially facilitated through my networked ethnographic methods.

Newsroom Fragments and Journalistic Ecosystems

What can we learn about the changes in journalism through an ethnographic study of a local news ecosystem? In this final section, I want to briefly highlight three of my research findings: the fragmenting of digital space, a cross-institutional overlap between various forms of newswork, and finally, the movement of news facts through a journalistic ecosystem. All of these findings were facilitated by an ethnographic focus that was ecosystemic rather than organizational or institutional and that drew on Actor-Network Theory as a way of thinking about socio-technical journalistic practice.

The Material Fragmentation of Digital Space

By chronicling the emergence of a digital news ecosystem over time, we can gain a deeper understanding of the material fragmentation of news space. Whereas powerful local news institutions—particularly newspapers—could once claim to represent the local news-consuming public as such, digital fragmentation makes such claims impossible. The change discussed here does not so much concern the reality of the local public—as a variety of post-Habermasian theorists have noted (Fraser, 1990; Warner, 2002), the public sphere has always been more diverse than simplified Enlightenment-era images of it would suggest. Rather, the change lies in the materiality of that public and in particular institutions that claim to represent it. Take coverage of local Philadelphia sports, to name one seemingly trivial example. Sport has always been an activity that unites particular geographic communities, and providing the news about professional sport has always played a key role in the daily business of local news organizations. And while residents of a community could always talk about sport amongst themselves—in bars, over the water cooler, at dinner with their friends, during the game itself—the material instantiation of that conversation was always the domain of the local news. Digital technologies have rendered some meaningful fraction of the once-fleeting conversations that took place around water-coolers, the kitchen table, or on emails lists, public for all the world to see. A well-known sports blogger with the Philadelphia Phillies blog "Beerleaguer" told me about his decision to become a blogger as part of a longer story about how he made public his formerly private thoughts about baseball:

> I'd heard the word blog before. Then, Google started offering a free service, called Blogger, and I helped put up a site for a friend. One day I found myself writing a long email about the Phillies, and I thought, maybe, instead of just sending this email to one person, or forwarding it around, I should just make an online home for this stuff. At first, I was just writing for myself. But then somebody links to you, and your stats go from three to six to fifteen. And then everybody is reading you. (interview, 6/12/2008)

One of the things I discovered during the course of my ethnographic research was just how profoundly these eruptions of public dialog into the communicative space once reserved for newspapers were affecting journalists' visions of themselves and their social roles. In their everyday conversations, newspaper reporters repeatedly reflected on their dawning knowledge that they were no longer the center of the public dialog in Philadelphia but simply a conversational actor amongst many. I heard much anguished conversation amongst reporters about the fact that newspapers were now competing with a range of bloggers and other online commentators. It was not so much that these bloggers and "citizen journalists" were inaccurate or unprofessional, though I certainly heard these criticisms. It had more to do with the fact that the once-insulated culture of the newsroom was suddenly forced to respond, or at least be aware of, a myriad of different voices. While newspaper executives were quick to point to their page view numbers and continued institutional strength as evidence that they remained the paper of record, lower-level reporters frequently bemoaned the fact that they were now only one voice in the larger Philadelphia conversation.

At the same time, when talking to bloggers and other online communicators, I was struck by the degree to which they inherently assumed they were only speaking to a small, niche community. In fact, they often positively compared themselves to newspaper journalists in this regard; "unlike the folks at the *Inquirer*," one citizen reporter told me, "we know who our audience is and aren't under any illusions we have to talk to everyone." The fact that I was able to compare different attitudes toward the online public at more traditional news organizations versus those at popular blogs was one of the advantages in my ecosystemic approach to news research. It was only in comparing these attitudes towards "speaking to everyone" versus "speaking to my niche" that I first began to notice the serious stress that the constant exposure to other online voices was causing more traditional reporters. By spending considerable time with members of the *emerging* media ecosystem, I gained new insights into the actions and anxieties of their traditional counterparts.

Varieties of Newswork

One of my primary research goals was to understand how journalistic work was changing in a digital age. The most comprehensive answers to this question usually came when I discovered that similar classes of newswork were coalescing at substantively different news organizations across the Philadelphia news ecosystem. In other words: it meant something when an editor at the radical citizen journalism website Philadelphia Indymedia and a Web producer at the much more traditional Philly.com repeated the same basic work processes and used the same news judgments. Major categories of work I noticed emerging across institutions included *reporting*—which had been a central form of work in newsrooms for nearly a century—and *news aggregation*, which was an older set of news practices being articulated in a new way. During my fieldwork, it was interesting to watch these two basic work practices adapt themselves to the organizations in which they were embedded, while at the same time percolating across disparate institutions in complex ways.

More often than not, aggregation and reporting were forms of newswork that traversed institutional boundaries. "Blogging," for instance—once seen as a fringe activity carried out by journalistic amateurs—has been actively embraced by traditional media organizations that now employ a variety of paid bloggers. We often think of blogging as somehow opposed to traditional journalism, or a form of newswork that undermines it in some way. Nevertheless, the blogging carried out at traditional news organizations were often more "reportorial" than aggregative and resembled the daily work of reporters operating under different technological circumstances and organizational imperatives. "I use my blog mostly as a portal for breaking news," a gossip blogger at the *Philadelphia Daily News* told me.

> There's very little difference between the way I do reporting and what I put on my blog. If I posted to [my blog] more often, if I linked more to other people, if I weighed in more on what *other* people were saying, it would be more like a blog.

The blogger at the *Daily News* summed up the attitude of most of the bloggers working at traditional newspapers when he told me that he didn't see himself as "wearing multiple hats": "I absolutely see everything I do as one assignment," he said. "And my job is to break news, stuff that's new, and distribute it. Blogging is in service to reporting, for me and for most of the bloggers here." In this way, the institutional importance of "reporting the news" could be understood, insofar as it asserted itself into the formerly fringe, opinion-laden practice of blogging.

"Aggregation," on the other hand, was a developing form of newswork. Web aggregators also operated inside traditional news organizations. During my fieldwork, I was struck by the manner in which the work of maintaining the homepage of the radical news website Philadelphia Indymedia was similar, from a process perspective, to the work of managing the homepage of the far more traditional Philly.com. Aggregators at both Philadelphia Indymedia and Philly.com spent most of their day ranking the news value of the different stories that were submitted to them, and then linking different stories together in larger news packages. The primary difference between aggregation and reporting was that the first form of newswork gathered and bundled information *from the Internet only*, rather than obtaining that information from non-Internet sources.

At Philly.com specifically, the primary role of the Web producer was to decide where a news story (usually written by a reporter at the *Inquirer* or the *Daily News*) should be placed on the Philly.com website. The work of news aggregation could be summarized as the inter-linking, bundling, and illustrating of Web content. The primary role of a Web producer was to coordinate amongst a series of quasi-institutionalized content producers. The tasks of the Web producers were thus to build links between independently produced news stories and to rank these bundled news stories according to a rapidly changing sense of its importance, popularity, and imagined newsworthiness.

Drawing parallels between these different varieties of journalistic labor was only possible because I spent a considerable amount of time with non-traditional journalists operating outside the dominant institutions of the local news ecosystem. Once again, the value of an ecosystemic approach to newsroom fieldwork was that it allowed me to comparatively analyze work practices that might have been invisible had I only spent time in traditional newsrooms.

The Movement of News

In an earlier media era one could characterize the movement of news as, in its simplest form, a movement from the social environment, to news institutions, to the media-consuming public. The most detailed analysis of the circulation of news might have characterized agenda-setting news institutions, like the *New York Times*, and those institutions that largely followed the lead of those primary institutions, like television news. But given the emergence of many quasi-news organizations, the existence of a deeply fragmented digital space, and the different varieties of newswork that are mutating across different organizational spheres, how do news stories diffuse today? To answer this question requires empirical analysis. Over a two-week period in the summer of 2008, I detailed how a simple story about the arrest and detention of four Philadelphia homeowners moved from activist websites, to the alternative press, to the inner pages of the *Philadelphia Daily News*, the city's daily tabloid newspaper, to major national blogs. Through a combination of network ethnography and more traditional, qualitative newsroom analysis, I tried to catalogue a step-by-step analysis of the circulation of this particular set of news facts, how they emerged, exploded, and then quickly faded away.

In this analysis of the "Francisville Four," I tried to detail the larger explanatory factors that might have contributed to the particular pattern of news diffusion I described, as well as the degree to which the factors I observed might be generalizable across other cases. I described a pattern of "iterative pyramiding," in which key websites positioned within highly particular communities of interest acted as bridges to larger, more diffused digital communities. I also argued that news movement in the particular case of the "Francisville Four" can be characterized by an unusual combination of fact entrepreneurship and what I called "categorical misrecognition." By fact entrepreneurship, I refer to the way that institutional and quasi-institutional actors took advantage of the new networked news ecosystem to advance stories, facts, or fragments of facts in a manner that would lead to maximum visibility. Intriguingly, both activists *and* traditional reporters often behaved like fact entrepreneurs during the unfolding of the Francisville Four story. At the same time, these different institutional actors misrecognized the identities and goals of other actors in the Philadelphia media sphere. Journalists judged activists and bloggers by journalistic criteria, for instance, while bloggers tended to embrace a form of news judgment that relied heavily on linking and context. A basic set of facts mutated as it spread across the ecosystem, and this mutation was driven by the organizational interests of various communicative actors. They sought to shape the story as it traveled.[1]

Without this ecosystemic approach to journalistic fieldwork, though, my efforts to model the origination and diffusion of news facts in our digital age would have been considerably more difficult. A heightened sensitivity to the work practices at less traditional news organizations—indeed, a sensitivity to their very existence—made me curious about the role that these organizations were playing in disrupting the former "linear" model of news diffusion. While the circulation of news can be, and has been, mapped using only content-analytical techniques (PEJ, 2010b), my presence in various newsrooms allowed me to activate my contacts simultaneously and gain insight into how nearly all the relevant news organizations were treating the story of the "Francisville Four." And this newsroom-based account of news diffusion, I would argue, adds considerable nuance to the emerging cluster of content-analytical news diffusion studies being carried out by various researchers.

Conclusion

The newsroom remains a central space in which the work practices, rhetoric, and technologies of journalism intersect to create an occupation. It has been the argument of this chapter, however, that this newsroom can no longer be viewed in isolation from the media ecosystem that surrounds it. I have tried to turn this somewhat philosophical claim (and philosophical conundrum) into a valid methodology for doing empirical ethnographic work, emphasizing the importance of networks in two senses—networks as quantifiable social networks, and networks as actor-networks engaged in the continual construction of the journalistic field. I have also tried to show that my particular empirical findings—particularly my understanding of how news diffuses, how work practices intersect across organizational boundary lines, and how the public sphere is fragmenting under the pressure of the explosion of digital content—were facilitated through my methodological choices.

The newsroom is not extinct. In many ways, it is more important than ever, for it remains, even now, a central locus in which a variety of fragmented actor-networks find themselves tied together to create an occupation. I have tried to distil in this chapter the paradox that only by "blowing up the newsroom" is it possible for us to understand the newsroom's importance in the study of journalistic work.

ENDNOTE

1 For more detail on the Francisville Four, see Anderson (2010).

Future Avenues for Research on Online News Production

Pablo J. Boczkowski

Common wisdom says that sequels are never as good as the original. But Chris Paterson and David Domingo have shown this does not necessarily have to be true by assembling an excellent collection of papers to follow up on the first iteration of *Making Online News: The Ethnography of New Media Production* (Paterson & Domingo, 2008). Taken together, these essays are a powerful indicator of how fertile and vibrant ethnographic study of online news production has become in a relatively short time given the paucity of field studies of online newsrooms existed a decade ago. They also provide a unique window into at least four directions in which this domain of inquiry has evolved in recent years—for a more extensive assessment of this scholarship at two points in time of its evolution, see Boczkowski (2002) and Mitchelstein and Boczkowski (2009).

First is the movement towards globalizing the space of data collection. From Zimbabwe to Norway, and from Belgium to the United States, the chapters in this volume present the reader with valuable evidence to start making sense of whether and how location matters in the evolution of online news. The findings from the two volumes edited by Paterson and Domingo, consistent with several books and articles, indicate a certain degree of homogeneity in practices and discourses across geographically dispersed locales. Second is the variety of news production settings studied. While the initial waves of scholarship on online news production tended to focus mostly on digital outlets of traditional print media, the previous chapters take the reader into digital outlets of wire agencies and broadcast media as well as online-only enterprises.

Third is the diversification in the focus of inquiry. These initial waves, strongly influenced by prior theorizing on news production, concentrated primarily on organizational and occupational issues. While still maintaining attention to these issues, the chapters assembled here have also broadened their foci to include comparatively less examined concepts such as discourse, categories, and networks. Fourth is the wide range of theoretical resources marshaled to interpret the findings, from traditional organizational sociology and mass communication ideas to the growing presence of concepts developed within the field of science and technology studies. This latter is perhaps the most striking theoretical innovation since, as I argued in *Digitizing the News: Innovation in Online Newspapers* (Boczkowski, 2004a), the sociology of news production largely overlooked the role of material culture in editorial work.

Given the trajectory in the recent evolution of this domain of inquiry illustrated by these chapters, what can we expect moving forward? What are the avenues of further development that seem potentially fruitful from today's vantage point? In the remainder of this Epilogue I take stock of what I learned from these chapters, and other recent scholarship, to propose four such avenues of possible intellectual renewal: shifting the stance of theoretical work from tributary to primary; broadening the analytical gaze from an exclusive focus on online news to more comparative perspectives that put this focus in the context of other social fields, organizations, and industries; expanding the methodological tool-kit from qualitative and ethnographic to also quantitative and mix-methods approaches; and, placing longstanding concerns with issues of news production in connection to analyses of the resulting news products and their appropriation and circulation in society.

As noted above, the chapters in this volume constitute a powerful indicator of the diversity of theoretical resources that have been brought to bear on this area of scholarship. Beneath this diversity, however, lies a somewhat limiting common denominator for theory-building efforts: the vast majority of the research has adopted a theoretically tributary or derivative stance, and therefore its conclusions have largely not been taken up by scholars who utilize similar theoretical resources in other areas of inquiry. That is, while the existing scholarship artfully draws upon a wide range of theoretical sources to make sense of the observed patterns in the data, it has comparatively made much less of an attempt to take advantage of the findings to contribute to the further development of these theoretical sources. For instance, utilizing the insights from science and technology studies for making sense of the role of technology in news production has proved to be extremely fruitful for some scholars who study online news making. But the fruitfulness of this approach has not resulted in major contributions to the further development of actor-network theory or the social construction of technology model, to name two popular sources of theoretical ideas in this volume. For instance, what could a scholar studying biotechnology from an actor-network theory perspective or a researcher making sense of nanotechnology from a social construction of technology vantage point learn about their respective objects of inquiry from reading the essays in this volume? A lot, I would argue, if the theoretical articulations that are relevant across objects of inquiry were made. But maximizing these learning opportunities would require a different approach to theory development, one that not only "draws upon" but also "gives back" to the sources of theoretical ideas and explicitly makes the effort to state the relevance of this theoretical work to scholars who study multiple objects of inquiry. In my opinion, this shift constitutes one of the most challenging, but also exciting, potential new directions of scholarship open to those who study the making of online news.

Another potentially important avenue for new developments in future work on making online news is a movement from concentrating the analytical gaze exclusively on issues having to do with online journalism to also including comparative analyses with other domains of work, social fields, and industries. As Eugenia Mitchelstein and I wrote in an essay about recent online news production scholarship, most research has been marked by "its relatively narrow focus on online news and the low level of attention it devotes to placing the empirical findings in the context of comparable processes in other industries. That is, making sense of the similarities and differences between innovation processes in online news and in other settings would help to ascertain what might be unique to the journalistic field and what might be shared across other domains of work" (Mitchelstein & Boczkowski, 2009, p. 576). Like most of the scholarship on the making of online news over the past fifteen years, the chapters in this volume do an important job in terms of illuminating the intricacies of producing content for the Web and its links to journalistic work in print and broadcast media. But what is missing from the analyses is an attempt to figure out what is different and what is shared between online news and other arenas of social action. Making sense of this difference and common ground would help broaden the audience of online news scholarship— which has largely been a sub-field among those academics concerned with journalism and new media—and appeal to researchers across communication and media studies, and also those beyond the boundaries of the field. Furthermore, combined with the previously stated need to shift theoretical work from tributary to primary, the exercise of broadening the analytical gaze could turn the study of online news into a critical intellectual space for re-thinking larger patterns concerning the role of information and symbolic work in contemporary digital culture.

A third area of possible new directions for scholarship on the making of online news centers on issues of methodological renewal. Heir to the traditions of the sociology of news production, the scholarship on online news making has relied almost exclusively on ethnographic approaches: scholars have gone inside the newsrooms where online news is made and combined observations and interviews with key actors involved in this process. The chapters in this volume follow along these lines, with relatively minor exceptions in the respective contributions of Anderson and Steensen. But these exceptions are quite telling. Blending social network analysis with ethnography enabled Anderson (see Chapter 12) to more precisely map the structure of inter-media conversations than what would have been possible by relying solely on qualitative techniques; and combining quantitative content analysis with interviews and observations allowed Steensen (see Chapter 7) to better ascertain the systematic effects of genre practices upon the resulting feature news stories. These two examples provide a glimpse into the analytical gains that could result from expanding the methodological tool-kit of (online) news production research—see also Singer (2008b). At least two paths could be pursued in the direction of mix-methods scholarship in this area. (These paths are not mutually exclusive.) First is undertaking quantitative analyses of data collected in the course of ethnographic research research (Quandt, 2008). For instance, in *News at Work: Imitation in an Age of Information Abundance* (Boczkowski, 2010), I conducted a quantitative analysis of field notes to ascertain the character and magnitude of differences between hard and soft news production. This not only enabled me to reach a level of precision that would have been impossible had I relied exclusively on qualitative techniques but also made some dimensions of these differences more salient (and others less so) than what appeared initially by using these techniques. Second is gathering complementary evidence that is primarily suited for quantitative analyses such as surveys, social network data, and content measures of news stories, among others. Anderson's and Steensen's chapters show the extent to which these additional data sources are not in detriment of the ethnographic evidence but actually enhance it by adding validity to the findings. In addition, they provide information that could not be obtained by using traditional ethnographic resources.

The fourth and final area for possible future development in scholarship about the making of online news alluded to above centers on breaking down the boundaries that still separate research on news production, the resulting products, and their appropriation by the public. Part of the legacy of the traditional sociology of news production was a strong focus on journalistic labor and a parallel comparative disregard for the systematic examination of the products of this labor and their reception and circulation in society. These divisions among production, product and consumption are, in turn, an artifact of deeper and longstanding divisions in the broader field of media and communication studies. Those who center on work, organizations and institutions have often proceeded independently from those who undertake textual analyses of media products and also from those who study communication effects. But, as a number of chapters in this volume begin to show, there are two main benefits that derive from breaking down these intellectual silos. First is the provision of a more realistic depiction of contemporary media phenomena, which are increasingly marked by a problematization of the boundaries between production and consumption. In light of current phenomena, these boundaries cannot be assumed beforehand in all cases but have to emerge as part of the inquiry process. Second is the development of a more robust theorizing on the links that tie production, products, and consumption. This, in turn, can shed light on each of these three stages in the media lifecycle in ways that could not be possible by focusing solely on one of them. Moreover, this type of integrative approach would contribute to a kind of intellectual renewal not just for online news scholarship, but also for the larger field of communication and media studies advocated above.

Taken together, these four directions for further research in the area of online news production would amount to a major transformation of this space of inquiry. But in light of the significant journey undertaken since this space started to coalesce a relatively short time ago, and the substantive evolution suggested by the chapters in this volume, it is not far fetched to imagine that when Paterson and Domingo put out a call for papers for the third volume of their anthology, most of this transformation might be already well in progress.

References

American Public Media (2010). *Marketplace FAQ*. Retrieved from http://marketplace.publicradio. org/about/faq.html

Andersen, S.S. (1997). *Case-studier og Generalisering*. Bergen: Fagbokforlaget.

Anderson, C.W. (2009). *Breaking Journalism Down: Work, Authority, and Networking Local News, 1997–2009*. Unpublished Doctoral Dissertation. Columbia University, New York.

Anderson, C.W. (2010). Journalistic Networks and the Diffusion of Local News: The Brief, Happy News Life of the 'Francisville Four.' *Political Communication*, 27(3), 289–309.

Associated Press. (2008). *A New Model for News: Studying the Deep Structure of Young-Adult News Consumption*. Retrieved from http://www.ap.org/newmodel.pdf

Baker, N. (2009). *Technology, Timeliness and Taste: The Battlefronts for the Twenty-First Century News Agency in International News Reporting: Frontlines and Deadlines*. J. Owen & H. Purdey (Eds.). Chichester, West Sussex: Wiley-Blackwell.

Banda, F., Mudhai, O.F., & Tettey, W.J. (2009). Introduction: New Media and Democracy in Africa—A Critical Interjection. In O.F. Mudhai, W.J. Tettey, & F. Banda (Eds.), *African Media and the Digital Public Sphere* (pp. 1–20). New York: Palgrave Macmillan.

Bardoel, J., & Deuze, M. (2001). Network Journalism: Converging Competences of Media Professionals and Professionalism. *Australian Journalism Review*, 23(2), 91–103.

Bechmann, A. (2009). *Crossmedia* (in Danish). PhD Thesis. Aarhus University, Denmark.

Bechmann Petersen, A. (2006). Internet and Crossmedia Productions: Case Studies of Two Major Danish Media Organizations. *International Journal of Emerging Technologies and Society, 4*(2), 84–107.

Berger, G. (2005). Powering African Newsrooms: Theorising How Southern African Journalists Make Use of ICTs for Newsgathering. In G. Berger (Ed.), *Doing Digital Journalism: How Southern African Newsgatherers Are Using ICTs* (pp. 1–14). Grahamstown: Highway Africa.

Bergstrom, A. (2008). The Reluctant Audience: Online Participation in the Swedish Journalistic Context, *The Westminster Papers in Communication and Culture, 5*(2), 60–80.

Berkenkotter, C., & Huckin, T.N. (1993). Rethinking Genre from a Sociocognitive Perspective. *Written Communication,* (10), 475–509.

Bertini, A. (2003). The BNC: AP's First Multimedia Newsroom, *AP World,* Winter, 2002/2003, Associated Press.

Bijker, W.E. (1995). *Of Bicycles, Bakelite and Bulbs: Towards a Theory of Socio-Technical Change.* Cambridge, MA: MIT Press.

Bird, S.E. (2010). Introduction, In S.E. Bird (Ed.), *The Anthropology of News and Journalism* (pp. 1–20). Bloomington, IN: Indiana University Press.

Boczkowski, P.J. (2002). The development and use of online newspapers: What research tells us and what we might want to know. In Lievrouw, L.A. and Livingstone, S. *Handbook of New Media.* London: Sage.

Boczkowski, P. (2004a). *Digitizing the News: Innovation in Online Newspapers.* Cambridge, MA: MIT Press.

Boczkowski, P.J. (2004b). The Processes of Adopting Multimedia and Interactivity in Three Online Newsrooms. *The Journal of Communication, 54*(2), 197–213.

Boczkowski, P.J. (2007). When More Media Equals Less News: Patterns of Content Homogenization in Argentina's Leading Print and Online Newspapers. *Political Communication, 24*(2), 167–180.

Boczkowski, P.J. (2009). Rethinking Hard and Soft News Production: From Common Ground to Divergent Paths. *Journal of Communication, 59*(1), 98–116.

Boczkowski, P.J. (2010). *News at Work: Imitation in an Age of Information Abundance.* Chicago, IL: University of Chicago Press.

Boczkowski, P.J., & Ferris, J.A. (2005). Multiple Media, Convergent Processes, and Divergent Products: Organizational Innovation in Digital Media Production at a European Firm. *Annals of the American Academy of Political and Social Science. 597*(1), 32–47.

Boczkowski, P., & Lievrouw, L.A. (2008). Bridging STS and Communication Studies: Scholarship on Media and Information Technologies. *The Handbook of Science and Technology Studies* (pp. 949–977), Cambridge, MA: MIT Press.

Bødker, S., & Petersen, A.B. (2007). Seeds of Crossmedia Production, *Journal of Computer Supported Cooperative Work, 16*:539–566.

Born, G. (2005). *Uncertain Vision.* New York: Vintage.

Bourdieu, P. (1993). *The Field of Cultural Production. Essays on Art and Literature.* Cambridge, UK: Polity.

Bourdieu, P. (1998 [1996]). *On Television.* New York: The New Free Press.

Bourdieu, P. (2005). The Political Field, the Social Science Field, and Journalistic Field. In Benson, Rod & Eric Neveu (Eds.): *Bourdieu and the Journalistic Field* (pp. 29–47). Cambridge, UK: Polity.

Bowker, G. (2007). Actor-Network Theory. In *Blackwell Encyclopedia of Sociology, Blackwell Reference Online*. Blackwell. Retrieved from http://www.blackwellreference.com/public/tocnode?id=g9781405124331_chunk_g978140512433317_ss1-9

Boyd-Barrett, O. (1980). *The International News Agencies*. London: Constable.

Boyd-Barrett, O. (2008/1998). 'Global' News Agencies. In H. Tumbler (Ed.), *Critical Concepts in Media and Cultural Studies* (Vol. IV, pp. 22–36). London: Sage.

Boyd-Barrett, O. (2010). National News Agencies in the Turbulent Era of the Internet. In O. Boyd-Barrett (Ed.), *News Agencies in the Turbulent Era of the Internet Catalonia* (pp. 9–34), Government of Catalonia, Presidential Department.

Boyd-Barrett, O., & Rantanen, T. (1998). Defining News: Contestation and Construction. In O. Boyd-Barrett & T. Rantanen (Eds.), *The Globalization of News* (pp. 173–176). London: Sage.

Boyd-Barrett, O., & Rantanen, T. (2004). News Agencies as News Sources: A Re-Evaluation. In C.A. Paterson & A. Sreberny (Eds.), *International News in the Twenty-First Century Eastleigh* (pp. 31–46). Published by John Libbey Pub. for University of Luton Press.

Boyer, D. (2010). Digital Expertise in Online Journalism (and Anthropology). *Anthropological Quarterly, 83*(1), 125–147.

Brannon, J. (2008). Multimedia Journalism in Three U.S. Newsrooms. In C. Patterson & D. Domingo (Eds.), *Making Online News* (pp. 99–112). New York: Peter Lang.

Bruns, A. (2008). *Blogs, Wikipedia, Second Life and Beyond*. New York: Peter Lang.

Burgess, J., & Green, J. (2009). *YouTube: Online Video and Participatory Culture*. Cambridge: Polity.

Carey, J. (1987). The Press and Public Discourse. *The Center Magazine, 20*, 4–32.

Castells, M. (1996). *The Rise of the Network Society*. Cambridge: Blackwell.

Castells, M. (2009). *Communication Power*. Oxford: Oxford University Press.

Chari, T. (2009). Ethical Challenges Facing Zimbabwean Media in the Context of the Internet. *Global Media Journal: African Edition, 3*(1), 1–34.

Clare, J. (1998). *Town Criers of the Global Village: News Production at Associated Press Television*. Unpublished Master's Dissertation. University of Leicester, England.

Coleman, E.G. (2010). Ethnographic Approaches to Digital Media. *Annual Review of Anthropology, 39*, 1–19.

Cortada, J. (2006). *The Digital Hand* (Vol. 2). London: Oxford University Press.

Cottle, S. (2000). New(s) Times: Towards a "Second Wave" of News Ethnography. *Communications: The European Journal of Communication Research* (25), 19–41.

Cottle, S. (2007). Ethnography and News Production: New (s) Developments in the Field. *Sociology Compass, 1*(1), 1–16.

Cottle, S., & Ashton, M. (1999). From BBC Newsroom to BBC Newscentre: On Changing Technology and Journalist Practices. *Convergence, 5*(3), 22–43.

Currah, A. (2009). *What's Happening to Our News?* Oxford: Reuters Institute for the Study of Journalism.

Curran, J. (2002). *Media and Power*. London: Routledge.

Deuze, M. (2003). The Web and Its Journalisms: Considering the Consequences of Different Types of News Media Online. *New Media & Society, 5*(2), 203–230.

Deuze, M. (2004). What Is Multimedia Journalism? *Journalism Studies, 5*(1), 139–152.

Deuze, M. (2007). *Media Work.* Cambridge, U.K.: Polity.

Deuze, M. (2008a). Toward a Sociology of Online News? In C. Paterson & D. Domingo (Eds.), *Making Online News: The Ethnography of New Media Production.* New York: Peter Lang.

Deuze, M. (2008b). The Professional Identity of Journalists in the Context of Convergence Culture. *Observatorio, 7,* 103–117.

Deuze, M., Bruns, A., & Neuberger, C. (2007). Preparing for an Age of Participatory News. *Journalism Practice, 1*(3), 322–338.

Dijk, T.A.v. (1997). The Study of Discourse. In T.A.v. Dijk (Ed.), *Discourse as Structure and Process* (pp. 1–34). London, Thousand Oaks, CA; New Delhi: Sage.

Domingo, D. (2006a). *Inventing Online Journalism: Development of the Internet as a News Medium in Four Catalan Newsrooms.* PhD Dissertation. Universitat Autònoma de Barcelona, Spain. Retrieved from http://www.tesisenxarxa.net/TESIS_UAB/AVAILABLE/TDX-1219106-153347//dd1de1.pdf

Domingo, D. (2006b). Material and Organizational Constraints in the Development of Online Journalism. In Leandros, N. (Ed.) *The Impact of Internet on the Mass Media in Europe.* Suffolk: Arima.

Domingo, D. (2008a). When Immediacy Rules: Online Journalism Models in Four Catalan Online Newsrooms. In C. Paterson & D. Domingo (Eds.) *Making Online News: The Ethnography of New Media Production.* New York: Peter Lang.

Domingo, D. (2008b). Interactivity in the Daily Routines of Online Newsrooms: Dealing with an Uncomfortable Myth. *Journal of Computer-Mediated Communication, 13*(3), 680–704.

Domingo, D. (2008c). Inventing Online Journalism: A Constructivist Approach to the Development of Online News. In C. Paterson & D. Domingo (Eds.), *Making Online News: An Ethnography of New Media Production* (pp. 15–28). New York: Peter Lang.

Domingo, D., Quandt, T., Heinonen, A., Paulussen, S., Singer, J.B., & Vujnovic, M. (2008). Participatory Journalism Practices in the Media and Beyond: An International Comparative Study of Initiatives in Online Newspapers. *Journalism Practice, 2*(3), 326–342.

Downes, E.J., & McMillan, S.J. (2000). Defining Interactivity: A qualitative Identification of Key Dimensions. *New Media & Society, 2*(2), 157–179.

Downie, L., & Schudson, M. (2009). The Reconstruction of American Journalism. *Columbia Journalism Review.* Retrieved from http://www.cjr.org/reconstruction/the_reconstruction_of_american.php

Duffy, M., Thorson, E., Lacy, S., Riffe, D. (2010) *PEJ Report on Online Community Journalism Sites—Phase II.* Washington, D.C.: Pew Research Center. Retrieved from http://www.stateofthemedia.org/2009/chapter%20pdfs/PewKnightreport%2008%20FINAL.pdf

Edmonds, R. (2006). Two part series: *The New Associated Press — A News Strategy to Fill the Gaps; The New Associated Press — Under Construction.* Published online March 27 and March 29. Poynter Institute. Retrieved from http://www.poynter.org/content/content_view.asp?id=98942

Eliasoph, N. (1988). Routines and the Making of Oppositional News. *Critical Studies in Mass Communication, 5*(4), 313–334.

Eliasoph, N. (1997). Routines and the Making of Oppositional News. In *Social Meanings of News* (pp. 230–254), London: Sage.

Emmett, A. (2009). Networking News: Traditional News Outlets Turn to Social Networking Web Sites in an Effort to Build Online Audiences. *American Journalism Review*, December-January 2009. Retrieved from: http://www.ajr.org/article.asp?id=4646

Engebretsen, M. (2006). Shallow and Static or Deep and Dynamic? Studying the State of Online Journalism in Scandinavia. *Nordicom Review*, *27*(1), 3–16.

Epstein, E.J. (1973). *News from Nowhere: Television and the News*. New York: Random House.

Erdal, I.J. (2008). *Cross-Media News Journalism. Institutional, Professional and Textual Strategies and Practices in Multi-Platform News Production*. PhD Dissertation. University of Oslo, Norway.

Fairclough, N. (1995). *Media Discourse*. London: Edward Arnold.

Finberg, H. I. (2002). *Convergence and the Corporate Boardroom. Changing the Industry's Cultural Template*, Poynter Online. Retrieved from http://www.poynter.org/content/content_view.asp?id=11696

Fishman, M. (1980). *Manufacturing the News*. Austin, TX: University of Texas Press.

Fraser, N. (1990). Rethinking the Public Sphere: A Contribution to the Critique of Actually Existing Democracy. *Social Text*, 56–80.

Friend, C., & Singer, J. (2007). *Online Journalism Ethics: Traditions and Transitions*. Armonk, NY: M. E. Sharpe.

Fulton, K. (1996). A Tour of Our Uncertain Future. *Columbia Journalism Review*, *34*(6): 9–26.

Gade, P.J., & Perry, E.L. (2003). Changing the Newsroom Culture: A Four-Year Case Study of Organizational Development at the *St. Louis Post-Dispatch. Journalism & Mass Communication Quarterly*, *80*(2), 327–347.

Galtung, J., & Ruge, M.H. (1965). The Structure of Foreign News: The Presentation of the Congo, Cuba and Cyprus Crises in Four Norwegian Newspapers. *M.H. Journal of Peace Research*, *2*(1), 64–90.

Gans, H. (1979). *Deciding What's News: A Study of CBS Evening News, NBC Nightly News, Newsweek and Time*. New York: Random House.

Garber, M. (2010, July 16). No, Seriously: What the Old Spice Ads Can Teach Us About News' Future. *Nieman Journalism Lab*. Retrieved from http://www.niemanlab.org/2010/07/no-seriously-what-the-old-spice-ads-can-teach-us-about-news-future/

García Avilés, J.A., & Carvajal, M. (2008). Integrated and Cross-Media Newsroom Convergence: Two Models of Multimedia News Production. The Cases of Novotécnica and La Verdad Multimedia in Spain. *Convergence*, *14*(2), 221–239.

García Avilés, J. Kaltenbrunner, A. Kraus, D. Meier, K., & Carvajal, M. (2008). *Newsroom Convergence. A Transnational Comparison*. Research report, Medienhaus Wien. Retrieved from http://www.medienhaus-wien.at/cgi-bin/file.pl?id=57

García-Avilés, J.A. Klaus Meier, J.A., Kaltenbrunner, A., Carvajal, M., & Kraus, D. (2009). Newsroom Integration in Austria, Spain and Germany. Models of media convergence. *Journalism Practice*, *3*(3), 285–303.

Gillmor, D. (2004). *We the Media. Grassroots Journalism by the People, for the People*. Sebastopol: O'Reilly. Retrieved from www.oreilly.com/catalog/wemedia/book/index.csp

Gilmour, D.A. & Quanbeck, A. (2010) Hegemony: Quiet Control over Convergence Textbook Content. *The Review of Communication*, 10(4), 324–341.

Gitlin, T. (1980). *The Whole World Is Watching*. Los Angeles, CA: University of California Press.

Glasgow University Media Group. (1980). *More Bad News*. London: Routledge.

Goffmann, E. (1959). *The Presentation of Self in Everyday Life*. New York: Penguin.

Hall, S., Critcher, C., Jefferson, T., Clarke J., & Roberts, B. (1978). *Policing the Crisis: Mugging, the State, and Law and Order*. London: Macmillan.

Harrison, J. (2010). UGC and Gatekeeping at the BBC. *Journalism Studies*, 11(2), 243–256.

Harrison, S., & Dourish, P. (1996). Re-Place-ing Space: The Roles of Place and Space in Collaborative Systems. *Proceedings of the 1996 ACM Conference on Computer Supported Cooperative Work* (pp. 67–76). Boston, MA; United States.

Hemmingway, E. (2004). The Silent Heart of News. *Space and Culture*, 7(4), 409.

Hemmingway, E. (2005). PDP, The News Production Network and the Transformation of News. *Convergence*, 11(3), 8.

Hemmingway, E. (2008). *Into the Newsroom: Exploring the Digital Production of Regional Television News*. London: Routledge.

Herman, E., & Chomsky, N. (1988). *Manufacturing Consent: The Political Economy of the Mass Media*. New York: Pantheon.

Hermida, A. (2009). The Blogging BBC: Journalism Blogs at 'the World's Most Trusted News Organisation.' *Journalism Practice*, 3(3), 268–284.

Hermida, A., & Thurman, N. (2008). A Clash of Cultures: The Integration of User-Generated Content Within Professional Journalistic Frameworks at British Newspaper Websites. *Journalism Practice*, 2(3), 343–356.

Howard, P. (2002). Network Ethnography and the Hypermedia Organization: New Media, New Organizations, New Methods. *New Media & Society*, 4(4), 551–575.

Huang, E., Davison, K., Shreve, S., Davis, T., Bettendorf, E., & Nair, A. (2006). Facing the Challenges of Convergence. Media Professionals' Concerns of Working across Media Platforms, *Convergence*, 12(1), 83–98.

Ibelema, M., & Powell, L. (2001). Cable Television News Viewed As Most Credible. *Newspaper Research Journal*, 22(1), 41–51.

IFJ/ILO. (2006). *The Changing Nature of Work. A Global Survey and Case Study of Atypical Work in the Media Industry*. Research report of the International Federation of Journalists (IFJ) and the International Labour Office (ILO). Retrieved from http://www.ifj.org/pdfs/ILOReport070606.pdf

Jarvis, J. (2009). *Product v. Process Journalism: The Myth of Perfection v. Beta Culture*. Retrieved from http://www.huffingtonpost.com/jeff-jarvis/product-v-process-journal_b_212325.html

Jenkins, H. (2006). *Convergence Culture: Where Old and New Media Collide*. New York: New York University Press.

Jensen, J.F. (1998). Interactivity: Tracking a New Concept in Media and Communication Studies. *Nordicom Review*, 1, 185–204.

Jones, B., & Paterson, C. (2010). Theorising Convergence in International News Production. Paper for the connect/reconnect/disconnect: ICS PhD Conference, University of Leeds.

Jorgensen, D. (1989). *Participant Observation: A Methodology for Human Studies*. Newbury Park, CA: Sage.

Karbasfrooshan, A. (2010, January 30). Context Is King: How Videos Are Found and Consumed Online. *TechCrunch*.

Karlsson, M. (2007). *Nätjournalistikk, En explorative fällstudie av digitale mediers karaktarsdrag på fyra nyhedssajter*, Lunds Universitetsforlag.

Kiss, J. (2007). Reuters' 'mojo' experiments with Nokia. *Guardian.co.uk*. Retrieved from: http://www.guardian.co.uk/media/pda/2007/oct/23/reutersmojoexperimentswith

Klinenberg, E. (2005). Convergence: News Production in a Digital Age. *The Annals of the American Academy of Political and Social Science*, 597(1), 48–64.

Kovach, B., & Rosenstiel, T. (2007). *The Elements of Journalism: What Newspeople Should Know and the People Should Expect*. New York: Three Rivers.

Kraut, R.E., Fussell, S.R., Brennan, S.E., & Siegel, J. (2002). Understanding Effects of Proximity on Collaboration: Implications for Technologies to Support Remote Collaborative Work. In P. Hinds & S. Kiesler (Eds.), *Distributed Work* (pp. 137–162). Cambridge, MA: MIT Press.

Krebs, V. (2008). Social Network Analysis. *org.net*. Retrieved from http://www.orgnet.com/sna.html

Kunelius, R. (2001). Conversation: A Metaphor and a Method for Better Journalism? *Journalism Studies*, 2(1), 31–54.

Küng, L. (2008). *When Innovation Fails to Disrupt. A Multi-lens Investigation of Successful Incumbent Response to Technological Discontinuity: The Launch of BBC News Online*. Jönköping: Media Management and Transformation Center, Jönköping International Business School.

Lagerkvist, J. (2008). Introduction: Why Ethnography? In C. Paterson & D. Domingo (Eds.), *Making Online News: The Ethnography of New Media Production*. (pp. 1–11). New York: Peter Lang.

Latour, B. (1986). The Power of Association. In J. Law (Ed.), *Power, Action and Belief*. London: Routledge.

Latour, B. (1987). *Science in Action*. Cambridge, MA: Harvard University Press.

Latour, B. (2005). *Reassembling the Social: An Introduction to Actor-Network Theory*. Oxford: Oxford University Press.

Lewis, J., Williams, A., & Franklin, B. (2008). Compromised Fourth Estate? UK News Journalism, Public Relations and News Sources. *Journalism Studies*, 9(1), 1-20.

Lewis, S.C., Kaufhold, K., & Lasorsa, D.L. (2010). Thinking about Citizen Journalism: The Philosophical and Practical Challenges of User-Generated Content for Community Newspapers. *Journalism Practice*, 4(2), 163–179.

Lindholm, M. (2008). Hva er en nettavis? In Ottesen R. & Krumsvik, A. *Journalistikk i en digital hverdag*. Kristiansand, Norway: IJ-Forlaget.

Lindlof, T.R., & Taylor, B.C. (2002). *Qualitative Communication Research Methods*. Thousand Oaks, CA: Sage.

Lowe, G.F. (2010). *The Public in Public Service Media*. Göteborg: Nordicom.

Luff, J. (2002). *Technology in Transition: Newsroom Automation Systems*. Retrieved from http://broadcastengineering.com/mag/broadcasting_technology_transition_newsroom/

Mabweazara, H.M. (2010). 'New' Technologies and Journalism Practice in Africa: Towards a Critical Sociological Approach. In N. Hyde-Clarke (Ed.), *The Citizen in Communication: Re-visiting Traditional, New and Community Media Practices in South Africa* (pp. 11–30). Capetown: Juta & Co.

Machill, M., & Beiler, M. (2009). The Importance of the Internet for Journalistic Research. *Journalism Studies, 10*(2), 178–203.

Magid, L. (1997). All Roads Lead to Reuters. Currents.net. Retrieved from www.currents.net/magazine/national/1513/uout1513.html

Mason, J. (2006). *Qualitative Researching*. London: Sage.

McChesney, R.W., & Nichols, J. (2010). *The Death and Life of American Journalism: The Media Revolution That Will Begin the World Again*. Philadelphia, PA: Nation.

McManus, J. (1994). *Market-Driven Journalism*. Thousand Oaks, CA: Sage.

McNair, B. (2003). From Control to Chaos: Towards a New Sociology of Journalism. *Media, Culture & Society, 25*, 547–555.

McNair, B. (2006). *Cultural Chaos: Journalism, News and Power in a Globalised World*. London: Routledge.

Meier, K. (2007). Innovations in Central European Newsrooms. Overview and Case Study. *Journalism Practice, 1*(1), 4–19.

Meyrowitz, J. (1985). *No Sense of Place: The Impact of Electronic Media on Social Behavior*. New York: Oxford University Press.

Miller, C. (1994). Genre as Social Action. In A. Freedman & P. Medway (Eds.), *Genre and the New Rhetoric* (pp. 23–42). London: Taylor & Francis.

Mitchelstein, E., & Boczkowski, P.J. (2009). Between Tradition and Change: A Review of Recent Research on Online News Production. *Journalism, 10*(5), 562–586.

Moyo, D. (2007). Alternative Media, Diasporas and the Mediation of the Zimbabwe Crisis. *Ecquid Novi: African Journalism Studies, 28*(1&2), 81–105.

Moyo, D. (2009). Citizen Journalism and the Parallel Market of Information in Zimbabwe's 2008 Elections. *Journalism Studies, 10*(4), 551–567.

Mukundu, R. (2006). Zimbabwe Independent. In G. Berger & F.M. Sibanda (Eds.), *What the Newsroom Knows: Managing Knowledge within African Newspapers* (pp. 28–38). Grahamstown: Highway Africa.

Northrop, K.J. (2005). *What to Do About Convergence in 2005?* IFRA Newsplex. Retrieved from http://www.ifra.com

Nyamnjoh, F. (2005). *Africa's Media: Democracy and the Politics of Belonging*. London: Zed.

Nygren, G. (2008). *Nyhetsfabriken: Journalistiska yrkesroller i en förändrad medievärld*. Lund: Studentlitteratur.

O'Sullivan, J., & Heinonen, A. (2008). Old Values, New Media. *Journalism Practice, 2*(3), 357–371.

Palmer, M. (1998). What Makes News. In O. Boyd-Barrett & T. Rantanen (Eds.), *Globalization of News* (pp. 177–190). London: Sage.

Paterson, C. (1997). Global Television News Services. In *Media in Global Context* A. Sreberny-Mohammadi, D. Winseck, J. McKenna and O. Boyd-Barrett (Eds.) London: Edward Arnold, 145-160.

Paterson, C. (2007). International News on the Internet: Why More Is Less. *Ethical Space: The International Journal of Communication Ethics, 4*(1), 57–66.

Paterson, C. (2010). The Hidden Role of Television News Agencies: 'Going Live' on 24 Hour News Channels. In S. Cushion, & J. Lewis (Eds.) *Has 24 Hour News Changed the World? The Global Impact of Rolling News*. New York: Peter Lang.

Paterson, C. (2011). *Television News Agencies: The World from London.* New York: Peter Lang

Paterson, C., & Domingo, D. (Eds.). (2008). *Making Online News: The Ethnography of New Media Production.* New York: Peter Lang.

Paterson, C., & Zoelner, A. (2010). The Efficacy of Professional Experience in the Ethnographic Investigation of Production. *Journal of Media Practice, 11*(2), 97–109.

Paulussen, S., & Ugille, P. (2008). User Generated Content in the Newsroom. Professional and Organisational Constraints on Participatory Journalism. *Westminster Papers in Communication and Culture (WPCC), 5*(2), 24–41.

Pavlik, J.V. (2001). *Journalism and New Media.* New York: Columbia University Press.

PEJ: Project for Excellence in Journalism. (2010a). *The State of the News Media 2010: An Annual Report on American Journalism.* Washington: Pew Project for Excellence in Journalism.

PEJ: Project for Excellence in Journalism. (2010b, January 11). *How News Happens: A Study of the News Ecosystem of One American City.* Journalism.org. Retrieved from http://www.journalism.org/analysis_report/how_news_happens

PEJ: Project for Excellence in Journalism. (2009). *The State of the News Media: An Annual Report on American Journalism.* Retrieved from http://www.stateofthemedia.org/2009/index.htm

Pfau, M., Holbert, L.R., Zubric, S., Pasha, N.H., & Lin, W. (2000). Role and Influence of Communication Modality in the Process of Resistance to Persuasion. *Media Psychology, 2,* 1–33.

Philo, G. (2007). Can Discourse Analysis Successfully Explain the Content of Media and Journalistic Practice? *Journalism Studies, 8*(2), 175–196.

Plesner, U. (2009). An Actor-Network Perspective on Changing Work Practices: Communication Technologies as Actants in Newswork. *Journalism, 10*(5), 604.

Pryor, L. (2002). Old Media Firms Dig a Grave with Shovelware. *Online Journalism Review.* Retrieved from http://www.ojr.org/ojr/technology/1017969861.php

Puijk, R. (2008) Ethnographic Media Production Research in a Digital Environment. In C. Paterson & D. Domingo (Eds.), *Making Online News* (pp. 29–43). New York: Peter Lang.

Purcell, K., Rainie, L., Mitchell, A., Rosenstiel, T., & Olmstead, K. (2010). *Understanding the Participatory News Consumer: How Internet and Cell Phone Users Have Turned News into a Social Experience.* Washington: Pew Research Center.

Quandt, T., & Singer, J. (2008). Convergence and Cross-Platform Journalism. In K. Wahl-Jorgensen & T. Hanitzsch (Eds.), Handbook of Journalism Studies (pp. 130–144). Oxford: Routledge.

Quandt, T., & Heinonen, A. (2009). *User-Generated Content as Challenge to Traditional Journalistic Ideology.* Paper presented to the 2009 ICA Conference "Keywords in Communication." Chicago, IL, May.

Quandt, T. (2008). News Tuning and Content Management: An Observation Study of Old and New Routines in German Online Newsrooms. In Paterson, C., & Domingo, D. (Eds.). (2008). *Making Online News: The Ethnography of New Media Production* (pp. 77–98). New York: Peter Lang.

Quinn, S. (2005). Convergence's Fundamental Question. *Journalism Studies, 6*(1), 29–38.

Reddick, R., & King, E. (1995). *The Online Journalist: Using the Internet and Other Electronic Resources.* Orlando, FL: Harcourt.

Schlesinger, P. (1978). *Putting Reality Together: BBC News.* London: Methuen.

Schudson, M. (2000). The Sociology of News Revisited. In J. Curran & M. Gurevitch (Eds.), *Mass Media and Society* (pp. 175–200). London: Arnold.

Schudson, M. (2003). *The Sociology of News*. New York: W. W. Norton.

Schmitz Weiss, A., & Domingo, D. (2010). Innovation Processes in Online Newsrooms as Actor-Networks and Communities of Practice. *New Media & Society, 12*(7), 1156–1171.

Schultz, I. (2006). *Bag om nyhedskriterierne—en etnografisk feltanalyse af nyhedsværdier i journalistisk praksis*, Skriftserie for journalistik at Roskilde Universitetscenter.

Shoemaker, P.J., & Reese, S. (1991). *Mediating the Message: Theories of Influences on Mass Media Content*. New York: Longman.

Silcock, B.W., & Keith, S. (2006). Translating the Tower of Babel? *Journalism Studies, 7*(4), 610–627.

Singer, J.B. (2004). Strange Bedfellows? The Diffusion of Convergence in Four News Organisations. *Journalism Studies, 5*(1), 3–18.

Singer, J.B. (2005). The Political J-Blogger: 'Normalising' a New Media Form to Fit Old Norms and Practices. *Journalism, 6*(2), 173–198.

Singer, J. (2008a). Five Ws and a H: Digital Challenges in Newspaper Newsrooms and Boardrooms. *International Journal on Media Management, 10*(3), 122–129.

Singer, J.B. (2008b). Ethnography of Newsroom Convergence. In Paterson, C., & Domingo, D. (Eds.). (2008). *Making Online News: The Ethnography of New Media Production* (pp. 157–170). New York: Peter Lang.

Singer, J.B. (2010). Quality Control. *Journalism Practice, 4*(2), 127–142.

Singer, J.B., Hermida, A., Domingo, D., Quandt, T., Heinonen, A., Paulussen, S., Reich, Z., & Vujnovic, M. (2011). *Participatory Journalism: Guarding Open Gates in Online Newspapers*. New York: Wiley-Blackwell.

Steen, J., Blombergsson, M., & Wiklander, J. (2005). Useful Buildings for Office Activities. *Facilities, 23*(3), 176–186. ABI/INORM Global.

Steensen, S. (2009a). Online Feature Journalism: A Clash of Discourses. *Journalism Practice, 3*(1), 13–29.

Steensen, S. (2009b). The Shaping of an Online Feature Journalist. *Journalism, 10*(5), 702–718.

Steensen, S. (2010). *Back to the Feature. Online Journalism as Innovation, Transformation and Practice*. PhD Dissertation. University of Oslo, Norway.

Steensen, S. (forthcoming). The Featurization of Journalism. What Feature Journalism is and how it Transforms as Genre. Article accepted for publication in *Nordicom Review*.

Strauss, A., & Corbin, J. (1998). *The Basics of Qualitative Research: Techniques and Procedures for Developing Grounded Theory*. London: Sage.

Strinati, D. (1995). *An Introduction to Theories of Popular Culture*. London; New York: Routledge.

Sullivan, J.L. (2004). *On the Rewards and Perils of "Studying Up": Practical Strategies for Qualitative Research on Media Organizations*. Paper presented at the IAMCR Annual Conference: Media Production Analysis Working Group Porto Alegre, Brazil.

Thurman, N. (2008). Forums for Citizen Journalists? Adoption of User Generated Content Initiatives by Online News Media. *New Media & Society, 10*(1), 139–157.

Tuchman, G. (1973) Making News by Doing Work: Routinizing the Unexpected. *The American Journal of Sociology, 79*(1), 110–131.

Tuchman, G. (1978). *Making News: A Study in the Construction of Reality*. New York: Free Press.

Turner, F. (2005). Actor-Networking the News. *Social Epistemology, 19*(4), 321–324.

Tushman, M., & Anderson, P. (1986). Technological Discontinuities and Organizational Environments. *Administrative Science Quarterly, 31*(3), 439–465.

Van Dijk, J.G.M. (2006). Digital Divide Research, Achievements and Shortcomings. *Poetics, 34*(4–5), 221–235.

Vartanova, E., & Frolova, T. (2010). News Agencies in Russia: Challenging Old Traditions and New Media. In O. Boyd-Barrett (Ed.), *News Agencies in the Turbulent Era of the Internet Catalonia* (pp. 224–246). Government of Catalonia, Presidential Department.

Vujnovic, M., Singer, J.B., Reich, Z., Quandt, T., Paulussen, S., Hermida, A., Heinonen, A., & Domingo, D. (2010). Exploring the Political-Economical Factors of Participatory Journalism. *Journalism Practice, 4*(3), 285–296.

Wahl-Jorgensen, K., Williams, A., & Wardle, C. (2010). Audience Views on User-Generated Content: Exploring the Value of News from the Bottom Up. *Northern Lights, 8*(1): 177–194.

WAN-IFRA. (2010). World Press Trends. Edition 2010. Darmstadt: WAN-IFRA.

Wardle, C., Williams, A., & Wahl-Jorgensen, K. (2008). *UGC @ the BBC: Understanding its Impact upon Contributors, Non-Contributors, and BBC News.* Report from a Knowledge Exchange Project Funded by the AHRC and the BBC.

Warner, M. (2002). Publics and Counterpublics. *Public Culture, 14*(1), 49.

Weaver, D.H., & Wilhoit, G.C. (1991). *The American Journalist: A Portrait of U.S. News People and Their Work* (2nd ed.). Bloomington, IN: Indiana University Press.

Weick, K. (1995). *Sensemaking in Organizations.* Thousand Oaks, CA: Sage.

Williams, A., Wahl-Jorgensen, K., & Wardle, C. (forthcoming). 'More Real and Less Packaged': Audience Discourses on Amateur News Content and Their Effects on Journalism Practice. In K. Andén-Papadopoulos & M. Pantti (Eds.), *Amateur Images and Global News*, Bristol, PA: Intellect.

Williams, A., Wardle, C., & Wahl-Jorgensen, K. (2010). Have They Got News for Us? Audience Revolution or Business as Usual at the BBC? *Journalism Practice.* iFirst. Retrieved from http://dx.doi.org/10.1080/17512781003670031

Willig, I. (before Schultz). (2007). The Journalistic Gut Feeling. Journalistic Doxa, News Habitus and Ortodox News Values. *Journalism Practice, 1*(2).

Winseck, D.R., & Pike, R.M. (2007). *Communication and Empire.* Durham, NC: Duke University Press.

Winston, B. (1998) *Media Technology and Society. A History: from the Telegraph to the Internet.* London, New York: Routledge.

WEF: World Editors Forum. (2008). *Trends in Newsrooms 2008.* Paris: WAN/WEF.

Xin, X. (2010). Xinhua News Agency in the Context of the "Crisis" of News Agencies. In O. Boyd-Barrett (Ed.), *News Agencies in the Turbulent Era of the Internet Catalonia* (pp. 248–265). Government of Catalonia, Presidential Department.

Xu, X. (2009). Development Journalism. In K. Wahl-Jorgensen & T. Hanitzsch (Eds.), *The Handbook of Journalism Studies* (pp. 357–370). New York: Routledge.

.

Contributors

C.W. Anderson is an assistant professor in the Department of Media Culture, College of Staten Island, CUNY (USA). He has held fellowships at the Harvard University Nieman Lab, Yale Law School, and the New America Foundation. His research focuses on the shifting ontologies of digital journalism. He blogs at cwanderson.org.

Anja Bechmann is an assistant professor of Information and Media Studies at Aarhus University (Denmark) and board member of the Centre for Internet Research (CFI). Her research focuses on cross-media as a digital transition in traditional media organizations and as digital cross-service strategies for Internet companies. Follow her at Twitter: @anjabp.

David Domingo is a senior lecturer in the Department of Communication Studies at Universitat Rovira i Virgili (Tarragona, Catalonia). He was visiting assistant professor at the University of Iowa in 2007–8. His research focuses on the adoption of innovations in online newsrooms. His weblog is dutopia.net.

Amira Firdaus is a PhD candidate at the University of Melbourne (Australia) and and a tutor at University of Malaya (Malaysia). Her thesis explores the intersection between new media and newswork at local and global news channels operating in Malaysia. She is also an associate editor and member of the founding editorial team of *Platform: Journal of Media and Communication*.

Davy Geens is online editor at the Belgian newspaper *De Standaard* (www.standaard.be). He was researcher at the Vrije Universiteit Brussel from 2006 until 2009.

Within the framework of a multidisciplinary project on Flemish E-publishing Trends (FLEET) he did research on the organizational dynamics of news production.

Hayes Mawindi Mabweazara is a lecturer in Journalism Studies at University College Falmouth, (England, UK). He has previously taught in Zimbabwe at the National University of Science and Technology and at the Zimbabwe Open University. His research focuses on the intersections between digital technologies and journalism practice in Africa and the wider global south.

Jannie Møller Hartley is a PhD candidate of the Department of Communication, Business and Information Technologies at Roskilde University (Denmark). Her research deals with online news practices, news values and journalistic self-understanding in Denmark and comparative content analysis of online news.

Chris Paterson is a senior lecturer at the University of Leeds' Institute of Communications Research (UK). His research concerns journalism and international communications. His book on international television news agencies was also published by Peter Lang in 2011.

Steve Paulussen is a senior researcher at the IBBT research group for Media & ICT at Ghent University (Belgium) and a lecturer in journalism and new media at the Vrije Universiteit Brussel and the University of Antwerp. He has published research on profiles of journalists, online journalism, newspaper innovation and youth and new media.

Sue Robinson is an assistant professor at the University of Wisconsin-Madison's School of Journalism & Mass Communication (USA). Her research focuses on how authority over information transforms in environments with new technologies such as the online newsroom.

Steen Steensen is an associate professor in the Department of Journalism, Library and Information Science at Oslo University College (Norway). His research interests include online journalism, feature journalism and the ideology of journalism. He blogs at steenyo.wordpress.com and contributes to the Online Journalism Blog: www.ojb.org.

Nikki Usher recently received her PhD from the University of Southern California's Annenberg School for Communication and Journalism. She is an assistant professor at George Washington University's School of Media and Public Affairs. Her work focuses on the transformation of journalism in the digital age.

Brooke Van Dam is an assistant professor of journalism at Azusa Pacific University (USA). She completed her PhD at City University London. Brooke's research looks at online journalists and their changing news production processes. She tweets at @brookevandam.

Kristel Vandenbrande worked for many years as lecturer and researcher at the Department of Media & Communication Studies of the Vrije Universiteit Brussel (Belgium).

Karin Wahl-Jorgensen is a reader at Cardiff University's School of Journalism, Media and Cultural Studies (UK). She is the author and editor of numerous books including *Journalists and the Public* (Hampton Press, 2007) and *Citizens or Consumers?* (Open University Press, 2005), and is currently co-writing a book about *Disasters and the Media* (Peter Lang, forthcoming).

Claire Wardle is a digital media consultant, trainer and researcher, specializing in journalism, social media and UGC. She currently works with the BBC College of Journalism (UK). She previously worked at the Cardiff School of Journalism, Media and Cultural Studies. She tweets as @cwardle.

Andy Williams is a lecturer at Cardiff University's School of Journalism, Media and Cultural Studies (UK). His research interests are media convergence and the rise of user-generated content and the role of PR in British media, particularly in science, health, and environment news. His Twitter persona is @llantwit.

Index

Digital Formations

General Editor: Steve Jones

Digital Formations is an essential source for critical, high-quality books on digital technologies and modern life. Volumes in the series break new ground by emphasizing multiple methodological and theoretical approaches to deeply probe the formation and reformation of lived experience as it is refracted through digital interaction. Digital Formations pushes forward our understanding of the intersections—and corresponding implications—between the digital technologies and everyday life. The series emphasizes critical studies in the context of emergent and existing digital technologies.

Other recent titles include:

Felicia Wu Song
Virtual Communities: Bowling Alone, Online Together

Edited by Sharon Kleinman
The Culture of Efficiency: Technology in Everyday Life

Edward Lee Lamoureux, Steven L. Baron, & Claire Stewart
Intellectual Property Law and Interactive Media: Free for a Fee

Edited by Adrienne Russell & Nabil Echchaibi
International Blogging: Identity, Politics and Networked Publics

Edited by Don Heider
Living Virtually: Researching New Worlds

Edited by Judith Burnett, Peter Senker & Kathy Walker
The Myths of Technology: Innovation and Inequality

Edited by Knut Lundby
Digital Storytelling, Mediatized Stories: Self-representations in New Media

Theresa M. Senft
Camgirls: Celebrity and Community in the Age of Social Networks

Edited by Chris Paterson & David Domingo
Making Online News: The Ethnography of New Media Production

To order other books in this series please contact our Customer Service Department:
(800) 770-LANG (within the US)
(212) 647-7706 (outside the US)
(212) 647-7707 FAX

To find out more about the series or browse a full list of titles, please visit our website:
WWW.PETERLANG.COM